Data-Driven Technologies and Artificial Intelligence in Supply Chain

This book highlights the importance of data-driven technologies and artificial intelligence in supply chain management. It covers important concepts such as enabling technologies in Industry 4.0, the impact of artificial intelligence, and data-driven technologies in supply chains.

- Provides solutions to solve complex supply chain management issues using artificial intelligence and data-driven technologies.
- Emphasizes the impact of a data-driven supply chain on quality management.
- Discusses applications of artificial intelligence, and data-driven technologies in the service industry, and the healthcare industry.
- Highlights the barriers to implementing artificial intelligence in small and medium enterprises.
- Presents a better understanding of the different risks associated with supply chains.

The book comprehensively discusses the applications of artificial intelligence and data-driven technologies in supply chain management for diverse fields such as service industries, manufacturing industries, and healthcare. It further covers the impact of artificial intelligence and data-driven technologies in managing the fast-moving consumer goods (FMCG) supply chain. It will be a valuable resource for senior undergraduate, graduate students, and academic researchers in diverse fields including electrical engineering, electronics and communications engineering, industrial engineering, manufacturing engineering, production engineering, and computer engineering.

Intelligent Data-Driven Systems and Artificial Intelligence
Series Editor: Harish Garg

Data-Driven Technologies and Artificial Intelligence in Supply Chain: Tools and Techniques
Mahesh Chand, Vineet Jain and Puneeta Ajmera

Data-Driven Technologies and Artificial Intelligence in Supply Chain

Tools and Techniques

Edited by
Mahesh Chand
Vineet Jain
Puneeta Ajmera

CRC Press
Taylor & Francis Group
Boca Raton London New York

CRC Press is an imprint of the
Taylor & Francis Group, an **informa** business

First edition published 2024
by CRC Press
2385 NW Executive Center Drive, Suite 320 Boca Raton, FL 33431

and by CRC Press
4 Park Square, Milton Park, Abingdon, Oxon, OX14 4RN

CRC Press is an imprint of Taylor & Francis Group, LLC

ISBN: 978-1-032-42673-0 (hbk)
ISBN: 978-1-032-61129-7 (pbk)
ISBN: 978-1-003-46216-3 (ebk)

DOI: 10.1201/9781003462163

Typeset in Sabon
by SPi Technologies India Pvt Ltd (Straive)

Contents

Preface

Supply chain management is undergoing a remarkable transformation in today's fast-paced and interconnected world. The emergence of data-driven technologies and artificial intelligence (AI) has transformed the way in which businesses operate, allowing them to optimize processes, increase efficiency, and provide superior client experiences. The purpose of this book, titled *"Data-Driven Technologies and Artificial Intelligence in Supply Chain: Tools and Techniques,"* is to provide a comprehensive overview of the intersection between supply chain management and the transformative potential of data-driven technologies and AI. It examines how these innovations are reshaping the entire supply chain, from sourcing and procurement to production, distribution, and customer service. At present, DDT and AI offer a more effective digital domain that promises to facilitate immediate access to information and effective decision-making in ever-increasing business environments. These DDT and AI technologies have enabled computers to perform tasks that were previously only performed by humans. Applications of DDT and AI into supply chain related-tasks holds high potential for boosting both top-line and bottom-line values.

This book is a practical guide for professionals, academics, and students who wish to comprehend and leverage the potential of data-driven technologies and AI in supply chain operations. It provides an in-depth examination of the fundamental concepts, methodologies, and practical applications that drive this revolutionary transformation. This book bridges the divide between theory and practice by combining theoretical insights with practical case studies and examples, enabling readers to comprehend the intricate workings of these advanced technologies and their impact on supply chain management in different fields.

We appreciate the experts and thought leaders who contributed their knowledge and insights to this publication. Their diverse perspectives and real-world experiences enrich the content and ensure its relevance to the swiftly evolving supply chain management landscape of the present day. As the editors of this book, we hope it serves as a valuable resource and reference guide for anyone attempting to navigate the complex landscape of DDT and AI in supply chain management. We invite readers to embark on this enlightening voyage by embracing innovation and leveraging the power of data to generate competitive advantage in the dynamic global marketplace.

Editors

Dr. Mahesh Chand is working as an Assistant Professor in the Mechanical Engineering Department at the J C Bose University of Science and Technology, YMCA Faridabad, India. He has passed his BE in Mechanical Engineering in 2007, ME in Production Engineering from Delhi Technological University (Formerly DCE), Delhi in 2009, and Ph.D. from YMCA University of Science and Technology, Faridabad, India. At the time of writing, he has some 13 years of teaching and research experience. He has completed a one-year project of "Visionary Learning Community of India (VLCI)" for manufacturing excellence in Indian SMEs in 2017–2018 (an Indo-German collaboration). He is a member of Indian Society of Technical Education (LMISTE), the Indian Institute of Industrial Engineering (LMIIIE), the Indian Science Congress Association (ISCA), IAENG International Association of Engineers, IAENG Society of Industrial Engineers and IAENG Society of Operations Research. He has one Indian and one Australian patent to his credit. He worked as a Guest Editor on the special issue "Operations and Supply Chain Management: Theoretical and Empirical Approaches" Administrative Sciences, 2018. At present, he is guiding 3 Ph.D. scholars in the field of supply chain management. He is working in the field of supply chain management, risk management, operation management, manufacturing technology and multi-criterion decision-making (MCDM) approaches. He has also published about 35 research papers in reputed international journals and conferences. He is teaching operations research, industrial engineering, supply chain management and numerical methods for undergraduate, postgraduate and research degree scholars.

Dr Vineet Jain is working as a Professor and Head of Department in the Department of Mechanical Engineering, Mewat Engineering College, Haryana Waqf Board, Government of Haryana, Nuh, Haryana-122107, India. He received the PhD degree in Mechanical Engineering at the Department of Mechanical Engineering, YMCA University of Science and Technology, Faridabad, India. He completed his B.E Mechanical with honours from N.I.T. Kurukshetra, India and M. Tech (Manufacturing and Automation) with honours from the YMCA Institute of Engineering, Faridabad, India. He is currently teaching about the internal combustion engine, inspection quality control and motor vehicle technology and has previously taught many other subjects, such as operation research, industrial engineering, and the basics of mechanical engineering. His areas of interest include manufacturing technology, operation research, industrial engineering, Industry 4.0, healthcare, Lean management, modelling and decision-making with expertise in ISM, GTMA, EFA, CFA, SPSS, AMOS, ANFIS, GA, TLBO and MADM, like AHP, TOPSIS, VIKOR, PROMETHEE, MOORA, PSI, CMBA, ELECTRE, WEDBA etc. At the time of writing, he possesses more than 22 years' experience in teaching and industry. He has written three books in mechanical engineering, which were published by h Dhanpat Rai Publication New Delhi, India. He has also published 36s paper in international journals from leading international publishers such as Elsevier, Springer, Emerald, Inderscience, Taylor & Francis, and Growing Science and has also presented papers at a number of national and international conferences. In addition, he is an Editorial Board Member of the *International Journal of Productivity and Performance Management* and the *International Journal of Data and Network Science*. He has also been a reviewer for 37 different international journals (indexed SCI/Scopus/ Web of Science), and has acted as a Technical Program Committee Member of 36 international conferences which were held overseas.

Dr Puneeta Ajmera is working as an Associate Professor in the Department of Public Health, School of Allied Health Sciences, Delhi Pharmaceutical Sciences and Research University, New Delhi. She has done her PhD in healthcare management. She has at present a total of 16 years' experience, which includes around 13 years of teaching and 3 years of industry experience. She is currently teaching research methodology, operations research and public health management and in the past has taught many subjects, such as biostatistics, women and child health, an introduction to the healthcare system, public health nutrition, and laws and ethics in public health. She has published more than 50 research papers in peer-reviewed international journals and has presented papers in national and international conferences. She has filed two patents and authored four books on healthcare system and health policy. Her areas of interest include maternal and child health, the globalization of healthcare, Industry 4.0 and Lean management in healthcare, decision-making in healthcare with expertise in MADM (AHP, TOPSIS, VIKOR) approaches and structural modelling techniques such as ISM and TISM.

Contributors

Gaurav Aggarwal
Mewat Engineering College
Nuh, India

Puneeta Ajmera
Delhi Pharmaceutical Sciences &
    Research University
New Delhi, India

Sayantan Chakraborty
Amity University
Gurgaon, India

Mahesh Chand
J C Bose University of Science &
    Technology, YMCA
Faridabad, India

Koustuv Dalal
Mid Sweden University
Sundsvall, Sweden

Pintu Das
National Institute of Pharmaceutical
    Education and Research
Guwahati, India

M. Kiruthiga Devi
Dr. M.G.R. Educational and
    Research Institute
Chennai, India

Rajiv Kumar Garg
National Institute of Technology
Jalandhar, India

M. Indiramma
B.M.S. College of Engineering
Bangalore, India

Vineet Jain
Mewat Engineering College
Nuh, India

Sheetal Kalra
Delhi Pharmaceutical Sciences &
    Research University
New Delhi, India

Dinesh Kumar
Protiviti Middle East Member
    Firm
Dubai, United Arab Emirates

J. Madhuri
Bangalore Institute of Technology
Bangalore, India

Jaseela Majeed
Delhi Pharmaceutical Sciences &
    Research University
New Delhi, India

Simpi Mehta
DPGITM
Gurgaon, India

Palka Mittal
Delhi Pharmaceutical Sciences &
    Research University
New Delhi, India

M. S. Murali Dhar
Vel Tech Rangarajan Dr. Sagunthala
    R&D Institute of Science and
    Technology Chennai
Chennai, India

N. Nagarathna
B.M.S. College of Engineering
Bangalore, India

Nittin
J C Bose University of Science &
    Technology, YMCA
Faridabad, India

Anish Sachdeva
National Institute of Technology
Jalandhar, India

Kartikay Saini
Scottish High International School
Gurugram, India

Luxita Sharma
Amity university
Gurgaon, India

Shivani Sharma
NGF College of Engineering
    &Technology
Palwal, India

Shweta Sharma
Govt. College for Girls
Gurugram, India

S. Shobana
R.M.K Engineering College
Chennai, India

Rajesh Kr Singh
Management Development Institute
Gurgaon, India

Sheetal Soda
National Institute of Technology
Jalandhar, India

Kuldeep Tomar
NGF College of Engineering
    &Technology
Palwal, India

M. Mujiya Ulkhaq
Diponegoro University
Kota Semarang, Indonesia

K. Unnamalai
Anna Adarsh College for women
Chennai, India

Laxmi Pandit Vishwakarma
Management Development
    Institute
Gurgaon, India

Ashok Yadav
National Institute of Technology
Jalandhar, India

Sapna Yadav
Delhi Pharmaceutical Sciences &
    Research University
New Delhi, India

Sheetal Yadav
Delhi Pharmaceutical Sciences &
    Research University
New Delhi, India

Chapter 1

A human-centered approach to artificial intelligence in the supply chain

M. Kiruthiga Devi
Dr. M.G.R. Educational and Research Institute, Chennai, India

M. S. Murali Dhar
Vel Tech Rangarajan Dr. Sagunthala R&D Institute of Science and Technology, Chennai, India

K. Unnamalai
Anna Adarsh College for women, Chennai, India

S. Shobana
R.M.K. Engineering College, Kavaraipettai, India

1.1 INTRODUCTION TO HUMAN-CENTERED AI

Personal computers (PCs) became popular in the 1980s, but program development mostly followed a "technology-centered strategy," which ignored the needs of ordinary users. The fields of human factors, computer science and psychology led to the development of human–computer interaction (HCI). The artificial intelligence (AI) research trend is evolving away from technology-oriented applications that focus on increasing output and performance and toward humanity-oriented applications that focus on augmenting human intellect with machine intelligence.

There are two ways to interpret HAI; one is AI under human control, as defined by Shneiderman (2020), and the other is AI on the human condition by Stanford HAI (2020). AI controlled by humans (or human-controlled AI) is an assessment that depends on the level of human control over it. Human-controlled AI only assists automation on one end, and AI-controlled autonomy exists on the other end. Shneiderman (2020) reports that human control AI is leveraged with AI automation and empowers human productivity while enhancing safety, efficiency, and trust (human control AI). The Stanford Harvard Artificial Intelligence Initiative 2020 (Stanford HAI) is an example of the reflective judgment of human condition.

DOI: 10.1201/9781003462163-1

The purpose of this type of AI is to augment human intelligence through machine intelligence by augmenting user understanding of computation and judgment processes, as well as adjusting algorithms through human context and societal phenomena to enhance the welfare of users. The characteristics of HCAI include human-like sensing ability, human-like self-executing ability, human-like cognitive abilities, and human-like self-adaptive ability to unpredictable environments.

1.1.1 Evolution of human-machine relationship

Prior to World War II, the human–machine connection was based on "people adapting to machines," but after World War II, it was based on "machines adapting to humans." Since the dawn of the computer era, the approach has evolved into human–computer interaction (HCI), which is the interaction between humans and non-AI computing systems. Automated machines (such as computer products) are often used as a tool to assist humans in monitoring and carrying out tasks (HCAI) as shown in Figure 1.1.

1.1.2 HCAI framework

The interactions between humans and artificial intelligence occur whenever a system is receiving human inputs, processing them, and then developing outputs; in other words, whenever a system does not yet attain full automation, or when it is not desired for it to do so. Considering a one-dimensional view of automation, readers were led to believe there would be a lower level of human control. However, Sheridan (1992, 2000) held strong beliefs

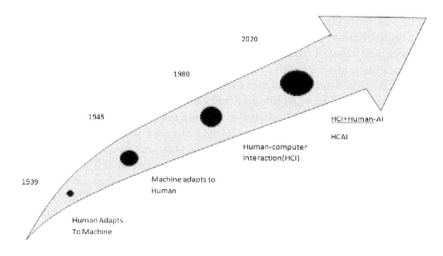

Figure 1.1 Evolution of human-machine interactions.

that supervisory control was essential, but Sheridan and Verplank (1978) were the first to formulate this one-dimensional vision. Two-dimensional HCAI (Shneiderman, 2020) clarifies how imaginative designers might envision highly automated systems while maintaining human control. HCAI (Shneiderman, 2020) illustrates how creative designers might design highly automated systems that can remain in control of the users. It separates control of the Human (y-axis) from computer automated (x-axis).

Digital cameras on mobile phones usually display an image of what a user would see if they pressed the big button. As users move their position or zoom in, the image updates smoothly. Simultaneously, the camera adjusts the aperture and focus automatically to compensate for trembling hands, a broad range of lighting situations (high dynamic range), and a variety of other factors. The camera's flash can be turned on or off, or it can be set automatically. Portrait, panorama, and video modes are available, as well as slow motion and time lapse. Users can also apply numerous filters, and after the image has been captured, they can make additional tweaks including brightness, contrast, cropping, and red eye removal.

These designs provide users with a lot of control while also allowing for a large degree of automation. Of course, mistakes happen, such as when the automated focus feature focuses on a nearby bush instead of the person standing directly behind it. To circumvent this error, competent users can touch the desired focus point, as shown in Figure 1.2.

The Prometheus Principles assist designers in creating understandable, predictable, and controlled interfaces for Real-Time Systems (RTS). Many of the ideas are spelled out in Human Interface Guidelines documents, such as that of Apple, which states that "User Control: … people, not apps, are in control" and "Flexibility: … (provide) users total, fine-grained control over their work." These and other guidelines which prioritize human control provide

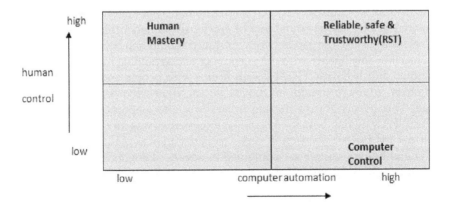

Figure 1.2 Illustrates human control over computer automation.

substantially different guidance than the naive pursuit of computer autonomy. These Prometheus Principles are a beginning point for achieving what most users, in my opinion, want: dependable, safe, and trustworthy systems. Research should be advocated toward sustainable AI to target the social impact of AI. Technology and humanity should be combined in order to understand this societal impact. Vinuesa et al. (2020) studied how AI can contribute to achieving sustainable development goals (SDGs) from two perspectives. Developing and using AI has helped achieve several SDG goals in a significant way as an enabler. However, it is biased because societal biases are underlying the integration of technology and humans in order to gain some understanding of this societal impact. Algorithms chase flawless results by repeated reinforcement learning, which is a determinant judgment. Because of the goal of perfection, these determining judgments frequently lead to extremes. It is thus simpler to establish biases in the algorithm and then steer or even grow erroneous biases, particularly in terms of ethnicity and gender. Explainable AI and interpretable machine learning (ML) are being developed by AI experts in the hopes of increasing the legitimacy of AI by allowing it to explain its conclusions.

One important aspect of HCAI is that intelligent systems should be built with a thorough understanding that they are components of a larger system that includes clients, operators, users, and other nearby stakeholders. Human-centered AI is a term used mostly by AI researchers and practitioners to describe intelligent systems that are designed with social responsibility in mind, meaning that such systems tend to be fair, accountable, interpretable, and transparent. HCAI is focused on enhancing human performance by amplifying, augmenting, and improving it, so that systems are safe, reliable, and trustworthy.

Human-centered AI focuses on algorithms that reside within a wider, human-based system, learning through human input and collaboration. Human-centered AI refers to systems that improve over time as a result of human input while also offering a positive human–robot interaction. Human-centered AI extends the boundaries of previously limited artificial intelligence solutions to bridge the gap between machines and human beings by developing machine intelligence with the purpose of understanding human language, emotion, and behavior. HCAI design begins with people and what is desirable from a human standpoint, taking into account what people desire and need. This means always putting others first and empathizing as a basic principle. Designing human–AI contact experiences was an essential subject that needed further research in order to better human–AI communication. Data and human science are combined in advanced contextual analytics to give particular behavioral data. Patterns emerge when analytics are applied to human actions and decisions. Contextual analytics combines data and human science to deliver radically better, more personalized customer experiences. When firms know exactly what their customers do and expect, they can build clear, educated business plans.

1.1.3 Human-centered design approach towards human-centered AI

The human-centered design approach was first established by Norman et al. (1999). Several ISO standards have been produced by the International Organization for Standardization (ISO) to standardize the strategy, process, and methodologies for human-centered design, including ISO 9241-210 and ISO 9241-220. Each method offers a unique perspective and enables for the investigation of specific areas of AI system design. Two components of work are highlighted in the strategy: technology and ethical design. They've learned that the next stage for AI development cannot simply be technology; it also has to be ethical and good to humans; AI is meant to supplement rather than replace human abilities. For example, in the absence of any effective interaction design mechanism, the design will be unable to ensure that humans remain the final decision-makers, will be unable to prevent possibly biased outcomes in human-centered ML, and will ultimately violate ethical design goals. Xu is working to close the gaps and overcome the specific obstacles that AI technology has brought with it, and he proposed a comprehensive approach for developing human-centered AI (HCAI); the HCAI approach includes three primary aspects of work interdependently: technology, human, and ethics, promoting the concept that we must keep human at the central position when developing AI systems. Shneiderman (2020) promotes HCAI by providing a two-dimensional design framework to guide the implementation of the HCAI approach in developing AI systems.

HUMAN: Researchers and developers of AI systems must identify effective usage scenarios for AI applications and implement human-centered HCI methods (e.g., research, modelling, interaction design, engineering) and other interdisciplinary approaches (e.g., modeling, interaction design, engineering, test). Developing AI systems that are helpful (in terms of value to humans and society) and usable (in terms of effective interaction design), with humans as the final decision-makers, is the goal of design.

ETHICS: In building AI systems, we must ensure fairness, justice, privacy, human decision-making authority, and accountability, as ethical considerations are far more critical in AI systems than in non-AI systems. The design goal is to create ethical and responsible AI by combining an interdisciplinary approach to HCI, meaningful human control, and existing techniques. The goal of ethical design will be reached by numerous disciplinary techniques such as system and HCI design, development practice, standards, organizational culture, & governance, and so on, from a macro perspective of sociotechnical systems.

Table 1.1 Comparison between human-centered design approaches and human-centered AI approaches (HCAI)

Physiology of design	Human-centered design approach	Human-centered AI approach
Focus	Non-AI computing systems' user experience	1. User needs, AI system utility (valued functionality), AI system usage scenarios and usability (user experience), intelligent interface technologies, and the ultimate decision-making authority of humans. 2. Human–machine hybrid intelligence, human intelligence augmentation, integrated AI and, augmenting people are all examples of technology. 3. Ethical AI design is done through meaningful interaction design, human control, and an interdisciplinary HCI approach, among other things.
Approaches and strategies for dealing with specific AI issues	HCI techniques (e.g., user needs, iterative UI prototyping and usability testing)	Human-centered/interactive machine learning methods, human-controlled autonomy, human–AI partnerships beyond interaction, and other HCI methods, as well as interdisciplinary methods. Additional HCI methods include AI as a design tool, longitudinal study, WOZ prototypes.
Multidisciplinary team	Computer science and HCI professionals	In addition to the traditional multidisciplinary team, additional professionals from the fields of AI, analytics, and neuroscience will be brought in.

TECHNOLOGY: Concentrate on the organic interaction of the three components. (a) Machine intelligence (AI) uses advanced and effective technologies such as models and algorithms, computing power, big data, and other methods to create a machine that resembles human intelligence to the greatest extent possible. (b) Human intelligence: Encourages the use of human intelligence augmentation (IA) technology, as well as the application of AI, psychology, biotechnology, neurotechnology, and other multidisciplinary methods to promote IA. (c) Human–machine hybrid intelligence: Promotes a strategy for AI technology development by integrating human functions and roles into human–machine systems and exploiting the complementary advantages of machine intelligence and human intelligence to create a more powerful form of human-centered hybrid intelligence as shown in Figure 1.3.

1.1.4 Modern supply chain management is driven by artificial intelligence and analytics

Supply chain and logistics using AI-based solutions is the process whereby intelligent computers may carry out problem-solving tasks. Without any

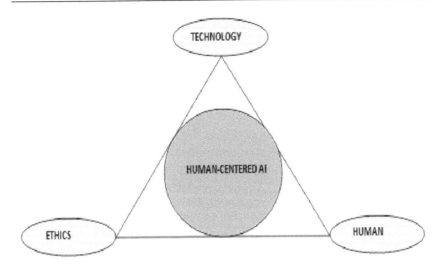

Figure 1.3 Dimensions of HCAI.

manual input, this automated process of smart industry manufacturing powered by the Industrial Internet of Things (IIoT) can operate the entire supply chain. AI integrated into the supply chain of business will increase the: Instrumented data comes from machine-like IoT devices with the aid of data analytics and modeling. When increased connectivity is used for enhanced decision-making through interconnection, intelligent assumptions will be more accurate and competent, as shown in Figure 1.4.

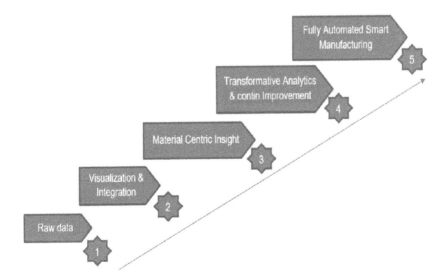

Figure 1.4 AI driving smart supply chain.

When there are vast volumes of data that may be used for forecasting, identifying efficiencies, and fostering innovation, supply chain data analysis aids in workflow optimization. This is known as supply chain analytics. To make wise data-driven decisions for your supply chain organization, there are primarily four types of supply chain analytics you may use. These are among the instances of supply chain analytics:

1. **Predictive:** This helps supply chain organizations to forecast future events and their commercial repercussions. For instance, this may involve reducing risks and disruptions with predictive analytics.
2. **Descriptive:** All types of internal and external data across the supply chain management are made visible and certain as a result.
3. **Prescriptive:** To maximize corporate value, this also entails working with logistical partners to cut time and effort. Supplier relationship management (SRM) is one such example of a prescriptive analytical technique.
4. **Cognitive:** The ideal application of this is in supply chain management to improve customer connections and experiences. To provide answers to complex problems, reports and dashboards analyze and use feedback data obtained through AI-driven systems.

With this sophisticated use of supply chain analytics, your company will automatically be able to explore ground-breaking concepts and meet the needs and demands of your customers. Our next topic of conversation is for you if you have not yet made up your mind about implementing AI and analytics for your company.

1.2 HUMAN-IN-THE-LOOP MACHINE LEARNING, REASONING AND PLANNING

Human-in-the-loop (HITL) is the process of combining machine and human intelligence to construct machine learning-based AI models. The goal of Human-in-the-Loop Machine Learning improves the accuracy of machine learning models, and means that human and machine intelligence can be combined to maximize accuracy, Faster reach the target accuracy for machine learning models, Increase efficiency by assisting human tasks with machine learning. The first category is to create human-in-the-loop AI systems at the system level, ensuring that humans are always included in AI systems. The system can be improved when a low level of confidence in the output causes humans to adjust input, thereby creating a feedback loop (Zheng et al., 2017; Zanzotto, 2019). Human-in-the-loop AI has recently been deployed for collaborative robots, including a human-input decision-making mechanism with ML and a knowledge base (Fu et al., 2019). The experimental results show that the robots performed better in dealing with

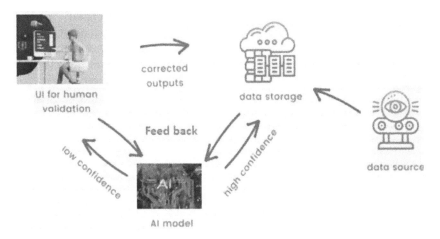

Figure 1.5 Depicts human in loop.

unstructured environments with the AI-human-in-the-loop solution rather than with the machine learning alone when the unstructured environment poses a high degree of uncertainty.

During the training and testing phases, it is essential to create models with higher accuracy faster and more efficiently, but it's also essential during the deployment phase. When AI models are deployed "in the wild", they are likely to come across situations for which they are unprepared because they are under-represented or mis-represented in their training data. In these cases, the models require human intervention so that humans can confirm the predictions of the AI model and provide feedback either to replace the AI-generated prediction or to be used to re-train and tune the model in the future.

The goal of humans in the loop is to improve and improve AI by modifying models and algorithms using human input and contribution. As previously stated, HITL can be used at several stages of the AI life cycle, as shown in Figure 1.5:

Training and testing: In order to speed up the learning process, humans in the loop can be involved during model training, validation, and testing. Humans can illustrate how tasks should be completed first, and then provide comments on the model's performance. This can be accomplished by correcting or evaluating the model's outputs, resulting in a reward function that can be utilized for reinforcement learning. When compared to typical supervised learning algorithms, learning via a combination of human demonstrations and evaluations has been shown to be faster and more sample-efficient.

Deployment: Human-in-the-loop procedures are especially crucial when training data is few, or when the data is unbalanced or incomplete, leaving us unsure whether or not the model is ready to handle all possible edge cases. Furthermore, even though the model often achieves high accuracy, human monitoring and double-checking may be required if model errors are likely to be extremely costly, as in circumstances where false negatives can cause irreversible damage, such as content moderation of user-generated content. The model can be connected to a labeling interface in either scenario, where outputs below a specific level of certainty are routed to be examined and validated by a person in real-time or in batches for future re-training.

As the world around us changes, AI models will need to be updated with current data to minimize model drift and detrimental biases, and people in the loop will be able to select and annotate new data to feed back into the model.

For a variety of reasons, a machine learning model may require human in-loop training

- There is no data that has been labeled. If there isn't one already, one must be made. To make one, utilize the Human in the Loop approach.
- The data set is constantly changing. If the environment that the data is supposed to represent changes rapidly, so must the model. With validation data sets reflecting current trends, human-in-the-loop learning can help keep models up to date.
- The data is difficult to label using automated methods. When unlabeled data is difficult to identify, human eyes are sometimes the only method to get the data classified.

Types of human-in-the-loop Machine Learning (HITL ML)

Human develop only model: Before deploying ML models, they may need to be trained first. If the goal is to create a model, you can create simulators that allow the model to make a prediction and show it to a person.

Human trains the model: When training an HITL model, it is assumed that the model's predictions will be poor at first, but that humans will be able to assess them. The goal is for the model to start performing at or better than human levels through human judgment. Simulators and labeling software come in a variety of shapes and sizes. Depending on the tasks, they can be either simple or highly complex. In software like spaCy's Prodigy or Label Studio, basic labeling tasks can be completed.

Human labels the data: Humans can be incorporated into the creation of a machine learning model through data labelling. Labelled data is required for any machine learning model. (Labels are already present in some datasets.) People are required to label data in HITL Machine Learning, and there is a lot of data to label.

1.2.1 Reasoning and planning

Planning with humans in the loop necessitates not just the ability to reason with ambiguity or incompleteness in terms of both environmental knowledge and human colleagues' preferences, but also the ability to deal with the various characteristics of human interaction. As robots grow more common in our daily lives, it's more crucial than ever to model acceptable or desired conduct in these autonomous creatures who share our space. There has been a lot of work done under the umbrella of "human-aware" planning, both in the context of path planning and in the context of task planning, that aims to provide social skills to robots so that they can produce plans that conform to desired behaviors when humans and robots work together. In reality, human-aware planning occupies a unique position in the field of multi-agent planning.

As we go from classical planning to multi-agent planning, we face issues in coordination, capability and commitment modeling, concurrency management, and so on. However, even within the multi-agent planning paradigm, adding a human to a multi-robot teaming scenario introduces new obstacles such as model completeness, priorities, and interaction issues. Furthermore, whether or not we can presume shared aims and expectations is determined by the presence or lack of a team, as shown in Figure 1.6. Multi-agent planning that includes the flavors of human–robot interaction but generally removes the assumptions commonly made in explicit teaming scenarios is referred to as human-aware planning. We'll look into these ideas with a live demonstration of a typical search and rescue scenario.

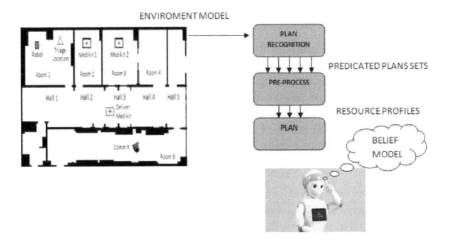

Figure 1.6 Illustrates human aware planning.

Resource Profiles: The first type of synergy we consider is passive synergy, in which the robot strives to avoid resource conflicts while trying to use the medkits – precisely, in the USAR context, the medkit is a constrained resource that is needed by both actors. We approach the problem by predicting and approximating resource usage probabilities over time using the human goal distribution and the known model of the human. As a result, the task of decoupling the human and robot plans essentially boils down to minimizing the resource usages produced by each of these plans. To elegantly minimize this overlapping, we turn the planning issue to an integer programming case. Remember that we don't have a team setting here, and that the human and robot aren't cooperating to avoid a dispute, therefore it's up to the robot to come up with a fail-safe plan that works for both of them.

Serendipity: The human does not expect the robot to assist him, since this isn't a team situation. As a result, any beneficial changes the robot makes to the world will look to the human as positive external events while he is carrying out his plan. For humans, such exceptions are referred to as fortuitous moments, and since the robot is attempting to induce such occasions, we refer to this planning paradigm as "planning for serendipity."

Interruptible: This defines the structure of the resulting joint human–robot strategy, making it compatible with the concept of fortuitous interventions.

Preservation constraints: These define a set of limitations on the structure of an interruptible plan that govern whether or not the robot is capable of producing a fortuitous intervention. We turn the planning problem into an integer programming instance using these two guiding principles, and illustrate how the robot can preempt help both with and without explicit communication.

1.2.1.1 Human Collaborators

Traditionally, mixed-initiative planning has focused on circumstances in which human specialists can contribute their complicated subject expertise to the planning process by evaluating computer planners. Surprisingly, a substantial subclass of human computation applications are those aimed at planning (e.g., tour planning) and scheduling (e.g., conference scheduling), where people are involved in the actual planning process. This might be thought of as a reverse mixed-initiative planning scenario, with humans ranging from a small group of professionals (for example, commanders making battle plans) to a vast audience. Explore numerous options for more complex automated planning in efficiently leading the crowd with a crude automated planner. The mismatch between the capacities of human workers and automated planners, however, makes straightforward application of existing planning technologies difficult. We identify and address two major obstacles that must be solved before such planning technology adaptation can take place: (i) understanding human employees' (and the requester's) inputs; and (ii) directing or criticizing human workers' plans, using only imperfect domain and preference models. To that aim, we devised a tour plan creation system that employs automated checks and alerts to improve the quality of plans made by humans.

1.2.1.2 Crowdsourced planning

Tour planning is used in crowdsourced planning to highlight the value of involving an expert. In a human planning task, an automated planner is used. Bear in mind that this is a situation when subject knowledge is vitally important. shallow, but we demonstrate that, even with such modest depths, We can't make plans with models, but we can make things with them. Constructive criticisms of the plan's proposed measures built.

- **Request:** This is the job description that explains what the job entails. It involves a high-level understanding of the objectives that must be met, and the preferences that must be fulfilled. The system decodes and then interprets it in order to come up with subgoals issue, needs to be addressed by the crowd. The audience has the option to As a response, perform the following two things:
  - **Adding New Activities:** To address unfulfilled subgoals or broken limitations, the community might add a new activity or action to the plan.
  - **Critiquing Existing Activities:** Existing activities can be criticized by the community if they believe they are invalid due to the limits or that there are better options available.

  During this process, the system checks for consistency in the background and provides ideas to improve the plan's quality, as well as computing the final plan using answer set programming as shown in Figure 1.7.

- **Proactive Decision Support:** The other end of the spectrum is collaborative planning, where domain experts collaborate with people who are also domain specialists. Note that this does not imply that the planner can replace people and develop the best plan on its own, because we don't know a priori what objective function the person is attempting to maximize. As in the previous scenario of crowdsourced planning, any assistance provided by the automated system would be in the form of suggestions to improve and optimize the human planning process.

  The workflow into two self-perpetuating components.

  - **Data-Driven Decision-Making:** In the real world there is a stream of information about the condition of the world that needs to be taken into consideration throughout the planning process in difficult domains. Humans can easily become overwhelmed by the

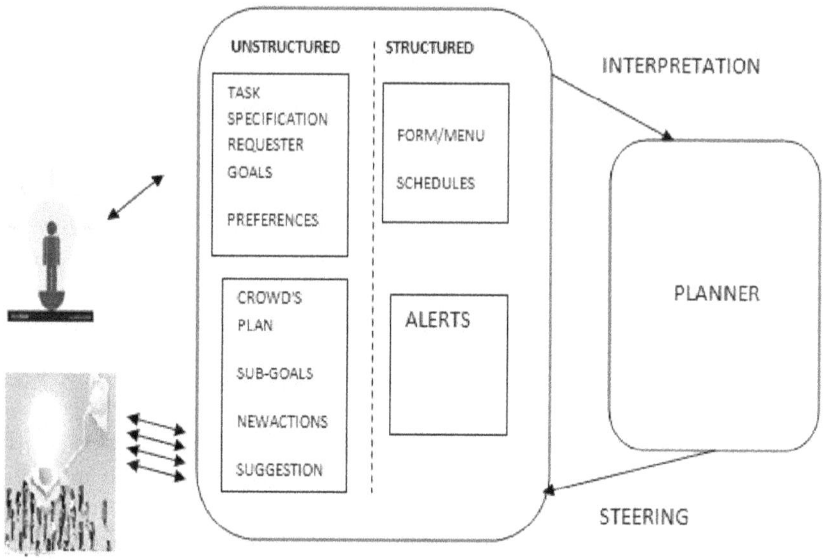

```
UNSTRUCTURED    STRUCTURED          INTERPRETATION

TASK
SPECIFICATION
REQUESTER       FORM/MENU
GOALS
                SCHEDULES
PREFERENCES

CROWD'S                             PLANNER
PLAN            ALERTS

SUB-GOALS

NEWACTIONS

SUGGESTION
                                    STEERING
```

Figure 1.7 Illustrates crowd source planning.

sheer volume of data coming from both structured (databases) and unstructured (Twitter) sources in this information age. As a result, while the planning component must be able to build and recognize plans using comprehensive domain knowledge, it must also take a data-centric approach.

– **Decision-Driven Data Gathering:** As the world develops, the planning process generates a demand for more data in future (which can thus be proactively predicted), completing the cycle of data required for planning and data generated and queried therewith. As a result, this process entails the detection and alignment of events, as well as the verification of data's trustworthiness and correctness from a variety of sources, as shown Figure 1.8.

1.2.2 Data analysis

Data analysis is the process of converting raw data into meaningful, appropriate information. With the advancement in artificial intelligence and machine learning we need to automate the data analysis process in order to resolve the issues related to data wrangling. This technology will enhance the speed and efficiency with which data can be transformed into meaningful information. Further, it will improve the productivity of data scientists and will be very beneficial to the society, industry and researchers. Major areas in which data analysis focuses are big data, social media, and customer and business analytics as shown in Figure 1.9.

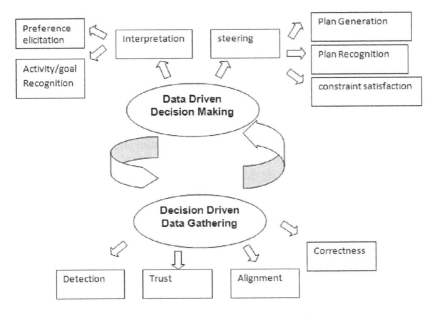

Figure 1.8 Depicts the data-driven decision-making and decision-driven data-gathering scenario.

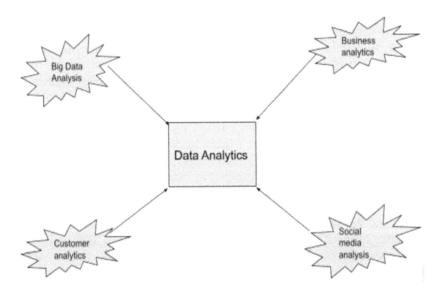

Figure 1.9 Data analytics focus area.

Data analysis learns from the raw data, finds patterns and makes decisions based on this. It also provides valuable suggestions with little involvement on the part of humans. The key objective of data analysis is to find meaning in data so that derived knowledge can be used to make proper decisions. The data analysis process includes many phases such as defining business problem statements, understanding and retrieving data, restoring data from multiple sources, cleaning data followed by feature extraction, then finding and removing abnormal data. Finally, there is the transformation and creation of data visualization and also of machine learning models.

AI-powered systems can sift through data from hundreds of sources and forecast what works and what doesn't. They can also delve into your consumers' data analytics and make predictions about their preferences, product development, and marketing channels.

Artificial intelligence (AI) and machine learning (ML) advancements have resulted in several achievements across a wide range of application disciplines. Complex systems are frequently created by merging the most recent advances in ML, interactive systems, data and visual analytics, and a variety of other domains. Human-centered machine learning is a new study topic that takes a holistic approach to ML< emphasizing the role of humans in the process. While artificial intelligence (AI) has resulted in significant advances in a variety of fields, understanding ML models remains a major difficulty. How can we make AI more accessible and interpretable, or, to put it another way, more human-centered, so that humans can comprehend and use these complicated models effectively? The study takes a human-centered approach to addressing these basic and practical AI difficulties by developing unique data visualization tools that are scalable, interactive, and simple to learn and use. Users can gain a better understanding of models by graphically exploring how huge input datasets affect models and their outcomes using such tools. It focuses on three interdependent parts, namely unified scalable interpretation, data-driven model auditing and learning complex models by experimentation. The first part provides scalable visual analytics tools to assist engineers in interpreting large-scale deep learning models at both the instance and subset levels. Secondly, design visual data exploration tools to support knowledge discovery by exploring data sets at different analysis stages, such as model comparison and fairness auditing; and finally, the construction of interactive tools that expands human access to learning complex learning models and browsing raw data sets.

By creating scalable, interactive visual analytics tools that allow users to explore and interact with ML models through data, users will have a better knowledge of machine learning models. New visualization tools, new data analytics paradigms, user interaction workflows, and scalable and accessible solutions all improve the human understanding of AI, speeding up its development, and boosting public confidence in this new technology. AI has various uses in today's society, and the rate of research is expanding to make human existence as simple as possible. To predict a user's search, most search engines employ artificial intelligence algorithms. Artificial intelligence is used by e-commerce companies to recommend products to users based on their recent searches. Data science is also used by delivery logistics organizations to increase their operational effectiveness. Huge amounts of data are retrieved via RSS Feeds and APIs to power websites such as Junglee, Trivago, and many others. Apart from the aforementioned few instances, data science and artificial intelligence are applied in every business that generates and analyses data.

Challenges of Data Analysis:

- Problem-solving and reasoning: Because there were so many alternative choices for an issue in the late 1980s and early 1990s, the algorithm employed to find solutions for huge reasoning problems was insufficient. As the problems grew larger and larger, this caused processing speeds to drop exponentially. The ideas of probability and economics were established by AI researchers as methods to deal with information uncertainty or incompleteness.
- Security: AI has been used to find and investigate security flaws. By using poor photos or any other means, classifiers can be readily tricked. Moving pictures or any other type of visual that is disturbed
- Manipulation and movement: When given a limited static environment, AI robots can readily detect and map their surroundings; but, when given a dynamic environment, or when mobility involves physical touch with the object, object recognition becomes harder to program.

1.2.3 Designing and prototyping

A designer can experience the real product without developing the actual product using a prototyping tool. It helps the designers to create a product at a faster pace and making the process more effective by allowing the user to experience the real feel of product. At present, the designers are keenly involved in understanding the technologies within the AI domains to prototype new services but prototype tools are still limited, which is a major concern in the area of machine learning. There are some tools, such as objectResponder, Wekinator, and Google Teachable, that allow designers to use ML for designing and prototyping. But for using these tools the designers require more technical skills or the process has to be simplified to have robust prototyping capability. Furthermore, tools are not readily available to designers. These restrictions make it difficult to create prototypes and to test machine learning systems and integrating rapid prototyping with machine learning.

But still some of the best AI-enabled prototyping tools which can be used for User Interface design and User Experience. Among these are Uizard,

Airbnb's AI, Balsamiq, InVision and Mockplus. Uizard is a popular tool which allows the machines to understand the GUI in the same way as humans do. It allows designers to generate code from sketches, generating mobile applications directly from sketches. Airbnb AI, developed by the American hospitality services company, is one of the prototyping tools which help designers to translate their idea directly from drawing board to reality.

This AI system recognizes the handmade design and automatically converts it into source code, thereby making the product development process simpler. One of the tools, Balsamiq, allows the designer to focus more on content than on colorful design layout. It's a UI design tool for creating wireframes to generate digital sketches from concepts or ideas. Another tool which is in widespread use is In Vision, which currently has over two million users across the world. This tool has proved very popular as it allows the designer to create clickable versions of designs in order to test and present it to the clients. In addition, it allows the designer to create animations, transition and to share the screen with their peers. Yet another app is Mockplus, one of the powerful tools in the AI design era. This makes communication easier, emphasizes creativity, and automates design and exporting design directly from sketches and Photoshop. It allows designers to build interactive prototypes.

Human-centered design (HCD) is a design method that places people and their needs, emotions, motivations, perspective, and behavior at the centre of the design process. People are not always "centered" when they are involved in the design process. Language frequently shows how people are regarded from various perspectives. People are frequently regarded as "human factors" in engineering, influencing the technology's performance. People are frequently referred to as "human resources" or "human capital" in management. HCD, by contrast, necessitates seeing humans as individuals. People having a variety of prior experiences, wants, goals, ambitions, hobbies, irrational decision-making, and lifestyles, all of which are rooted in unique cultural contexts. HCD is a paradigm shift in which humans are seen as integral to every aspect of the design rather than as an afterthought.

In the creation of computer systems such as AI, the mentioned methodologies were chosen to provide multiple views such as society, diversity, interaction, and human needs. Each HCD technique allows researchers to investigate and assess the impact of AI on people from a different perspective. The next sections of this chapter go through the implications and value of numerous HCD methodologies in examining the effects of AI systems on people:

- **Human-Centered Systems**
  Human-Centered Systems is the first HCD approach. This HCD technique enables researchers to investigate changes in large-scale social structures as a result of computer system design, implementation, and use, such as AI (Sawyer, 2005). Technology is dependent on the political,

intellectual, and cultural assumptions that give rise to it, according to this human-centered perspective. The impact of AI systems on social organization can be investigated using this HCD study technique. Pee, Pan, and Cui (2019), for example, looked into the joint knowledge work of an AI robotic system and people in a hospital environment. They discovered various types of knowledge embodiment, as well as the effects of embodiment on human–AI robotic system social connections. Such research allows for the identification of effects caused by the AI system's design and use, such as various and sometimes contradictory effects, benefits to some groups over others, moral and ethical implications, and reciprocal links with the broader social context (Sawyer, 2005). The designers' ideologies and cultural context in the design of the work system are another important part of Human-Centered Systems. Ideologies such as Taylor's (1907) scientific management, according to Cooley (1980), will affect how individuals are handled within the designed system. This is addressed in the next HCD method.

- **Social Design**

  According to Cooley, Social Design tackles the designers' ideological problem (1980). It's an HCD design approach that emphasizes the designer's social responsibility. As Papanek emphasizes, this approach incorporates the designer's duties in the design choices that have an impact on society (1973). The designer is the translator of social needs into AI system design. Examining the social and economic structure, as well as the function and intent of designers, allows for the identification of underlying values and motivations for certain AI solutions.

  The aim, decision-making, and responsibilities of designers in the development and deployment of AI systems can all be investigated using this HCD approach to AI research. This study looks into the socioeconomic factors that influence designers' decisions. For example, Facebook's targeted advertising business model influences the design and deployment of AI that promotes user engagement by screening content, which has influenced democratic elections. AI systems used on social media platforms have ethical implications that must be explored and resolved. Examining whether AI systems and business models are intended to replace or augment people reveals information about the designer's values and role in society, as shown in Figure 1.10. Examining

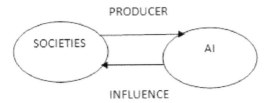

Figure 1.10 Social design.

the designer's ingrained socioeconomic and political processes, as well as the intention choices made, might reveal underlying motive and ideals. Designers must understand their function, philosophy, and socioeconomic processes in order to create AI systems that are good to society. Participatory Design is one of the HCD approaches that can transcend the designer's solitary viewpoint.

- **Participatory Design**
  The goal of Participatory Design is to democratize participation in system design (Bodker, 1996). This HCD method tackles issues of human impact such as democracy, power, and control. It permits collaboration with a variety of stakeholders. In the design of AI systems, participatory design is critical. This HCD research technique contributes by informing AI developers and cyberneticians about subtle differences across subgroups in society that could benefit from improved functional access to computer systems and adoption (Neuhauser & Kreps, 2011).

  Through a variety of ways, participatory design can assist in the creation of AI System ideas, as shown in Figure 1.11. However, the design space is only represented for the duration of the project, and users must have a fundamental understanding of what AI can and cannot achieve. Real-world participation in participatory design is generally limited since participants are often only involved while the project is running, power issues are not addressed as management takes final decisions, and participation is resource-intensive and frequently uncompensated (Bodker, 1996). Participatory Design takes into account the viewpoints of those involved in the project. Inclusive Design is another HCD technique that follows a similar approach.

- **Inclusive Design**
  Inclusive Design (Waller, Bradley, Hosking, & Clarkson, 2015) is a design strategy that includes and considers the requirements and behavior of diverse groups in order to make mainstream products, services, and systems accessible, usable, and beneficial for as many people as possible. This approach illustrates how varied populations are incorporated into the building of AI systems. This HCD study method analyses the usability, accessibility, and utility of AI design for certain categories of users. Inclusive computer system designs improve accessibility and

INTENT AND ROLE OF DESIGNER

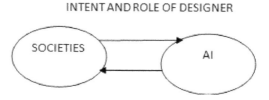

Figure 1.11 Participatory design.

usefulness while also addressing ethical concerns (Abascal & Nicolle, 2005). Individual differences, such as physiological, psychological, and sociological aspects, must be addressed to accommodate differences in the design (Benyon, Crerar, & Wilkinson, 2000). In AI systems, the decision of who to include has ethical concerns. When AI systems are trained, for example, ignoring specific groups of people results in racism and gender prejudice (Leavy, 2018). Machine biases in AI system design can be avoided by forming diverse design teams. It is not simply a matter of having more diverse and diversity-minded design teams when it comes to building AI systems that overcome biases; AI systems may also help identify gender and racial biases (Daugherty, Wilson, & Michelman, 2019). Early in the design process, HCD experiments that include and include different groups help discover and make biases evident. Interactions between individuals and AI systems are examples of such trials, as shown in Figure 1.12.

- **Interaction Design**
  Bill Moggridge and Bill Verplank were the first to suggest interaction design (Moggridge, 2007). This approach examines people's behavior, behaviors, and cognitive processes inside interactions to better understand and develop human–machine interactions (Norman, 1986, 2013; Norman & Draper, 1986). The interaction between humans and AI systems. This HCD study allows for prototyping of human–AI system interactions (Houde & Hill, 1997). In human–AI interaction research, a reverse Turing Test-like prototype known as Wizard of Oz is a common study tool. In AI interaction research, such as autonomous driving interactions (e.g., Fu et al., 2019; Rothenbucher, Li, Sirkin, Mok, & Ju, 2016) or human–robot interaction (e.g., Rothenbucher et al., 2016), prototyping was employed (e.g., Martelaro, Nneji, Ju, & Hinds, 2016; Shibata, Tashima, Arao, & Tanie, 1999). In the interface with the AI system, human behavior, activities, and emotions are all immediately observed. These findings allow researchers to investigate an AI system's immediate impact on individuals. Similarly, according to Xu (2019), HCI researchers can help with ethical AI design and technological advancement. This HCD method allows for the detection of

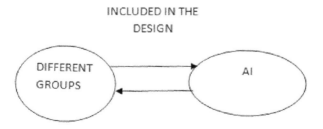

Figure 1.12 Inclusive design.

potentially hazardous interactions, as well as the creation of valuable experiences for people. According to Thaler and Sunstein, it can prod people into making better financial decisions (2008). Through technology such as AI, the same method is used to persuade humans to perform the desired behavior, as shown in Figure 1.13.

- **Persuasive Technology**
  Persuasive Technology is a method of attempting to change human attitudes, behavior, or both through the use of technology (Fogg, 2003). The underlying persuasive ability in human-computer interactions In the creation of AI systems, this HCD research strategy has two directions. Companies exploit AI systems' persuasive abilities to "hook" customers into purchasing extremely profitable products. The other is that this HCD research increases people's understanding of persuasive computers, allowing them to embrace such technologies to improve their own lives and detect when they are being persuaded by technology. Researchers looked into the persuasive abilities of AI systems like robotic agents and ambient intelligence (e.g., Midden & Ham, 2008; Verbeek, 2009). Berdichevsky and Neuenschwander (1999) proposed a paradigm for thinking about and preventing the ethical abuse of persuasive technology. This HCD study approach to AI systems includes an examination of AI system design as well as people's behavioral changes in everyday life.

- **Human-Centered Computing**
  A move away from human–machine interactions to the design of "interspaces," as described by Winograd (1997), is Human-Centered Computing. Interspaces incorporate people's lifestyles and the system design as present in everyday life of people (Hallnäs & Redström, 2002). In this approach, intelligence is viewed as an attribute of the combination of human– machine context (Ford, Hayes, Glymour, & Allen, 2015).

  Researchers looked into the persuasive abilities of AI systems like robotic agents and ambient intelligence (e.g., Midden & Ham, 2008; Verbeek, 2009). Berdichevsky and Neuenschwander (1999) proposed a paradigm for thinking about and preventing ethical abuse of persuasive technology. However, understanding how motivational tactics

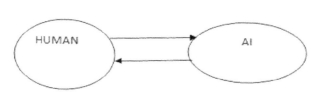

Figure 1.13 Interaction design.

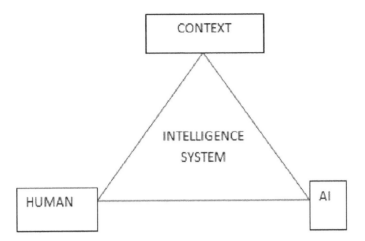

Figure 1.14 Illustrates human-centered computing.

function is important to discover genuine motivational techniques used in AI technology (Orji & Moffatt, 2018). This HCD study approach to AI systems includes an examination of AI system design as well as people's behavioral changes in everyday life, as shown in Figure 1.14.

- **Need Design Response**

  The Need-Design Response is the ultimate HCD technique, and it is based on McKim's work (1959, 1980). Every design, according to McKim (1959), is a response to a human need, which is frequently triggered by the natural and cultural environment or context. Every significant AI system design, from this perspective, must meet a need in a given context. This HCD research method enables you to identify needs and respond with a meaningful design (McKim, 1959). In the usage of created systems such as AI, the technique allows finding the underlying causes of people's motivation as outlined by Maslow (1987). Martelaro and Ju (2017) employed a Wizard of Oz prototype to discover needs when dealing with an AI system in real-world situations. In AI, the link between need and design shows moral consequences. The following criteria are used to evaluate this relationship: Is the "AI solution" that satisfies a "human demand" in a "context" potentially beneficial or harmful, as shown in Figure 1.15. Such an AI system might easily encourage people to engage in risky financial conduct. Addressing emotional needs, in particular, may be extremely profitable and requires a high level of morality, as not all human needs are good (McKim, 1959). This HCD research approach allows for the explicit expression of fundamental needs that the AI system design answers. When AI systems are able to model socio-cultural specific expectations and the behavior of individuals in order to predict human requirements, this method becomes particularly relevant (Riedl, 2019).

CONTEXT

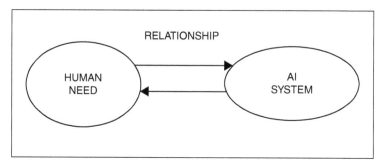

Figure 1.15 Need-design response technique.

1.3 USE CASES FOR AI AND ANALYTICS TO REDUCE SUPPLY CHAIN INTERRUPTION

It is time for contemporary supply chain businesses to equip themselves with trustworthy, automatic data visual analytics solutions. To minimise the interruption to the supply chain and maximise the profitability of your company, you can adhere to the principles of AI and analytics that are outlined below.

1.3.1 Warehouse supply and demand management with demand forecasting

In supply chain management, demand forecasting refers to the process of organising or anticipating the demand for materials to make sure you can provide the appropriate goods in the appropriate amounts to meet customer demand without producing a surplus. A surplus can be produced as a result of a forecasting inaccuracy, which is inefficient and expensive.

1.3.1.1 Need for demand forecasting

Since demand forecasting is the method used to develop the strategic and operational strategies, it is crucial to the supply chain. Consider it the beginning point for the majority of supply chain operations, including manufacturing, purchasing, inbound logistics, cash flow, and raw material planning. It also serves as the underlying hypothesis for strategic business activities.

Additionally, demand forecasting supports crucial company processes, including budgeting, scheduling manufacturing, evaluating risks, and buying raw materials. Above all, forecast accuracy helps merchants avoid both stock outs and overstocking while also reducing costs, boosting operational effectiveness, and enhancing customer satisfaction.

1.3.1.2 Working

Although a lot of information is provided very fast, sales forecasting really only involves using historical sales data to predict future consumer demand. Both qualitative and quantitative forecasting, which employ many sources and data sets to infer useful sales data, can be divided under the process. When historical sales information is available for a certain product and there is a known demand, the quantitative forecasting method is employed. It necessitates the application of mathematical formulas and data sets, including sales, revenue, and website analytics statistics. The qualitative approach, on the other hand, is based on the forecasters' intuition and experience, as well as developing technologies, shifts in pricing and availability, product lifecycles, and product upgrades.

1.3.1.3 Techniques

There are various techniques you can employ to forecast demand which fall under the categories of qualitative and quantitative forecasting:

- Collective Opinion collects historical data about consumer demand by utilising the expertise and experience of a company's sales force.
- Client survey methods offer important details about customer needs, wants, and expectations. Although it can be challenging to foresee actual demand, this data is useful for establishing sales forecasts.
- The Barometric Method, which makes use of economic indicators to gauge current activity and forecast developments.
- The Expert Opinion Method, which is seeking out professional judgment from outside consultants to determine future action.
- The market experiment method uses controlled market experiments to provide retailers with information about consumer behavior.
- The statistical method enables a business to establish performance history across time, identify patterns, and estimate potential future trends. It also enables a business to find and analyze the links between various factors.

1.3.1.4 Demand forecasting

All organizations can benefit from demand forecasting, but e-commerce brands and retailers can benefit the most since precise forecasts can help with inventory management and enhance the consumer experience. However, mastering a complex task such as forecasting accuracy for an online store is no easy feat. There are certain tried-and-tested methods that can simplify the procedure.

- **Collect the Right Data:** Ensure the appropriate type of data to make wise business decisions in order for your demand forecast to be successful. Focus on the statistics that provide you with the knowledge you need to make decisions, such as price patterns and the volume of visitors to your sales channels during a specific period of time. Avoid concentrating all of your data collection efforts on one specific product line. Focusing on the items and groups that bring in the most money and have the greatest consumer appeal is preferable.

- **Variable Adjust:** The daily interactions that effect sales data are influenced by a variety of factors. You must take into account all potential factors, such as unanticipated store closings or natural disasters, for your demand projection to be accurate. Another consideration is whether the product is in demand only occasionally or will continue to do so in the future. This can make it more difficult to generate an accurate projection.
- **Documents Sales and Demand Trends:** Regardless of the metric you select, you'll need a repeatable data analysis process that clearly shows whether the forecast is improving or deteriorating, identifies the areas that require the most improvement, assesses accuracy at the time of procurement, and provides accurate data by customer, branch, brand, product, and category.
- **Budget, Purchase and Allocate:** The only thing left to do is use the data you've gathered to develop a plan for how, where, and when to distribute your resources and purchase efforts once your demand forecast is in place. AI and machine learning for the sustainability of logistics and transportation.

1.3.2 AI and machine learning for the sustainability of logistics and transportation

To provide real-time insights into the durability of transport vehicles, IoT device data is generated from in-transit supply chain vehicles. Based on historical data, the ML systems built into the vehicles prescribe maintenance and predict failure. This will give you the opportunity to remove transient cars from the chain before the performance problem results in any kind of delivery delay. Not to mention the decrease in downtime, which can prevent serious mechanical breakdowns.

1.3.3 AI improves supply chain loading process portability

Detailed study of every aspect of the supply chain is required, including the loading and unloading of shipments and items from shipping containers. To find the most effective ways to load and unload the goods onto the containers, precision data modelling and AI are required. For direct, real-time insight into the loading process, modern supply chain firms combine software (SRM), hardware, and supply chain data analytics. The information gathered can also be used to create efficient and rapid process protocols for managing packages.

1.3.4 AI for supply chain cost reduction and revenue increase

The most efficient ways to negotiate better shipping and procurement rates, identify changes in the supply chain profit process, and manage courier contracts are AI and analytics tools. You can evaluate a standardized database that incorporates practically all supply chain components to provide financial decision-making. In general, the application of artificial intelligence in supply chain management is opening the door for fresh ideas in which platforms are standard for data mining and analysis of standards that can be used to generate money at a low cost. According to a Bloomberg study, the supply chain's overall cost has dropped to 12% over the past two years, resulting in profits.

1.3.5 Supply chain strategic sourcing based on data analytics

The identification of crucial suppliers and strategic partners is the most underappreciated application of AI and analytics in the supply chain. Standardize less expensive options and base supply performance metrics for compliance in this way. Using descriptive and predictive analytics also helps manufacturing sectors such as high-tech, Consumer Packaged Goods (CPG), and consumer electronics grow even more. Advanced self-service AI and analytics are essential for enabling a high level of independence and transparency in the supply chain company.

1.4 SUMMARY AND CONCLUSION

While AI technology has benefited many people, it still has the potential to harm humans if not properly developed. "Technology enhancement within the application development process and incorporation of a human-centered approach" is how the current third wave of AI might be described. The third wave of AI not only introduces new problems in terms of incorporating numerous "human factors" while designing AI systems, but also necessitates a more comprehensive human-centered approach, which presents a significant opportunity for the HCI community. In terms of human–computer interaction, AI systems have distinct properties that distinguish them from non-AI computing systems. AI systems can demonstrate distinct machine behavior and evolve to achieve human-like cognitive, self-executing, and symbiotic abilities. The various ways in which human-in-the-loop reasoning and planning differs from traditional planning will be expanded upon in the future.

This chapter also gives an overview of various human-centered design techniques and how they contributed to the creation of Human-Centered AI.

Artificial intelligence has numerous advantages for the supply chain management. After being accepted by businesses of all shapes and sizes, AI has become a standard tool for the supply chain. Every supply chain business model needs to be carefully connected with AI and analytics tools for optimization given the current situation. It is therefore imperative that your company employ AI development services if your supply chain firm is still lacking the current technologies outlined above. The HCAI framework differentiates human control from computer automation and clarifies that a good design is helpful to achieve a powerful combination of human control and automation. It also specifies when computer control is required for quick automated action. The HCD methodology described shows that human-centered AI requires people, including the designer, user, and other stakeholders, as well as their philosophies, practices, activities, interactions, and demands. To develop a truly human-centered AI strategy, multidisciplinary research combining domains such as Supply Chain, psychology, cognitive science, computer science, engineering, business management, law, and design is necessary.

REFERENCES

Abascal, J., & Nicolle, C. (2005). Moving towards Inclusive Design Guidelines for Socially and Ethically Aware HCI. *Interacting with Computers*, 17. DOI: 10.1016/j.intcom.2005.03.002

Benyon, D., Crerar, A., & Wilkinson, S. (2000). Individual Differences and Inclusive Design: Concepts, Methods, and Tools. In *User Interfaces for All*. DOI: 10.1201/9780429285059-2

Berdichevsky, D., & Neuenschwander, E. (1999). Toward an Ethics of Persuasive Technology toward an Ethics of Persuasive Technology. *Communications of the ACM*, 42(5), 51–58.

Bødker, S. (1996). Creating Conditions for Participation: Conflicts and Resources in Systems Development. *Human-Computer Interaction*, 11, 215–236. DOI: 10.1207/s15327051hci1103_2

Cooley, M. (1980). Computerization Taylor's Latest Disguise. *Economic and Industrial Democracy*, 1(4), 523–539. https://doi.org/10.1177/0143831X8014004

Daugherty, P. R., Wilson, H. J., & Michelman, P. (2019). Revisiting the Jobs Artificial Intelligence Will Create. *MIT Sloan Management Review*, 60(4), 0_1–0_8.

Fogg, B. J. (1998). Persuasive Computers: Perspectives and Research Directions. In *Proceedings of CHI 98* (pp. 225–232), New York, NY, ACM.

Fogg, B. J. (2003). Persuasive Technology: Using Computers to Change What We Think and Do. *Ubiquity*. 2022, Article no 5. https://doi.org/10.1145/764008.763957

Ford, K. M., Hayes, P. J., Glymour, C., & Allen, J. (2015). Cognitive Orthoses: Toward Human-Centered AI, *AI Magazine*, 36(4), 5–8. DOI: 10.1609/aimag.v36i4.2629

Hallnäs, L., & Redström, J. (2002). From Use to Presence – On The Expressions And Aesthetics Of Everyday Computational Things. *ACM Transactions on Computer-Human Interaction*, 9, 106–124. DOI: 10.1145/543434.543441

Houde, S., & Hill, C. (1997). Chapter 16 - What do Prototypes Prototype? In M. G. Helander, T. K. Landauer, & P. V. Prabhu (Eds.), *Handbook of Human-Computer Interaction* (Second Edition, pp. 367–381), North-Holland, ISBN 9780444818621. https://doi.org/10.1016/B978-044481862-1.50082-0

Leavy, P. (2018). Introduction to Arts-Based Research. In P. Leavy (Ed.), *Handbook of Arts-Based Research* (pp. 3–21). New York, NY: The Guilford Press.

Martelaro, N., & Ju, W. (2017). The Needfinding Machine. In *Proceedings of the Companion of the 2017 ACM/IEEE International Conference on Human-Robot Interaction* (pp. 355–356). DOI: 10.1145/3029798.3034811

Martelaro, N., Nneji, V. C., Ju, W., & Hinds, P. (2016). Tell Me More Designing HRI to Encourage More Trust, Disclosure, and Companionship. In *2016 11th ACM/IEEE International Conference on Human-Robot Interaction (HRI)* (pp. 181–188), Christchurch, New Zealand. DOI: 10.1109/HRI.2016.7451750

Maslow, A. H. (1987). *Motivation and Personality* (R. Frager, J. Fadiman, C. McReynolds, & R. Cox, Eds.), (3rd Edition), Boston: Addison Wesley.

McKim, R. H. (1959). Designing for the Whole Man. In J. E. Arnold (Ed.), *Creative Engineering Seminar*. Stanford, CA: Stanford, University.

McKim, R. H. (1980). *Experiences in Visual Thinking*. Belmont: Brooks/Cole Publishing Company.

Midden, C. J. H., & Ham, J. (2008). The Persuasive Effects of Positive and Negative Social Feedback from an Embodied Agent on Energy Conservation Behavior. In *2008 Symposium on Persuasive Technology* (pp. 9–13). AISB.

Moggridge, B. (2007). *Designing Interactions*. Cambridge: The MIT Press.

Neuhauser, L., & Kreps, G. (2011). Participatory Design and Artificial Intelligence: Strategies to Improve Health Communication for Diverse Audiences. AAAI Spring Symposium - Technical Report.

Norman, D., & Nielsen, J. (2013). 10 Usability Heuristics for User Interface Design. http://www.nngroup.com/articles/ten-usability-heuristics/

Norman, D. A., & Draper, S. (1986). User Centered System Design: New Perspectives on Human-Computer Interaction Lawrence Erlbaum Associates.

Norman, P., Conner, M., & Bell, R. (1999). The Theory of Planned Behavior and Smoking Cessation. *Health Psychology*, 18, 89–94. https://doi.org/10.1037/0278-6133.18.1.89

Orji, R., & Moffatt, K. (2018). Persuasive Technology for Health and Wellness: State-of-the-Art and Emerging Trends. *Health Informatics Journal*, 24(1), 66–91. DOI: 10.1177/1460458216650979

Papanek, G. F. (1973). Aid, Foreign Private Investment, Savings, and Growth in Less Developed Countries. *Journal of Political Economy*, 81, 120–130. http://dx.doi.org/10.1086/260009

Riedl, M. (2019). Human-Centered Artificial Intelligence and Machine Learning. *Human Behavior and Emerging Technologies*, 1, e117. DOI: 10.1002/hbe2.117

Rothenbucher, D., Li, J., Sirkin, D., Mok, B., & Ju, W. (2016). Ghost Driver: A Field Study Investigating the Interaction between Pedestrians and Driverless Vehicles. In *2016 25th IEEE International Symposium on Robot and Human Interactive Communication (RO-MAN)* (pp. 795–802), New York, NY, USA. DOI: 10.1109/ROMAN.2016.7745210

Sawyer, K. R. (2005). *Social Emergence. Societies as Complex Systems*. Cambridge University Press. https://doi.org/10.1017/CBO9780511734892

Sheridan, T. B., & Verplank, W. (1978). *Human and Computer Control of Undersea Teleoperators*. Cambridge, MA: Man-Machine Systems Laboratory, Department of Mechanical Engineering, MIT.

Shibata, T., Tashima, T., Arao, M., & K. Tanie (1999). Interpretation in Physical Interaction between Human and Artificial Emotional Creature. In *8th IEEE International Workshop on Robot and Human Interaction. RO-MAN '99 (Cat. No.99TH8483)* (pp. 29–34), Pisa, Italy. DOI: 10.1109/ROMAN.1999.900306

Shneiderman, B. (2020). Human-Centered Artificial Intelligence: Three Fresh Ideas. *AIS Transactions on Human-Computer Interaction*, 12(3), 109–124. https://doi.org/10.17705/1thci.00131

Thaler, R., & Sunstein, C. (2008). *Nudge: Improving Decisions about Health, Wealth, and Happiness* (p. 293). New Haven: Yale University Press, ISBN: 978-0-300-12223-7.

Verbeek, P.-P. (2009). Ambient Intelligence and Persuasive Technology: The Blurring Boundaries between Human and Technology. *NanoEthics*, 3(3), 231–242.

Waller, S., Bradley, M., Hosking, I., & Clarkson, P. J. (2015). Making the Case for Inclusive Design. *Applied Ergonomics*, 46(Part B), 297–303, ISSN 0003-6870. https://doi.org/10.1016/j.apergo.2013.03.012

Winograd, T. 1997. *The Design of Interaction. Beyond Calculation: The Next Fifty Years* (pp. 149–161). USA: Copernicus.

Wrege, C. D. (1995). F.W. Taylor's Lecture on Management, 4th June 1907: An Introduction. *Journal of Management History (Archive)*, 1(1), 4–7. https://doi.org/10.1108/13552529510082796

Xu, W. (2019). Toward Human-Centered AI: A Perspective from Human-Computer Interaction. *Interactions*, 26(4), 42–46. https://doi.org/10.1145/3328485

Zanzotto, F. M. (2019). Viewpoint: Human-in-the-Loop Artificial Intelligence. *Journal of Artificial Intelligence Research*, 64, 243–252. DOI: 10.1613/jair.1.11345

Zhang, D., Pee, L. G., & Cui, L. (2021). Artificial Intelligence in E-Commerce Fulfillment: A Case Study of Resource Orchestration at Alibaba's Smart Warehouse. *International Journal of Information Management*, 57, 102304, ISSN 0268-4012. https://doi.org/10.1016/j.ijinfomgt.2020.102304

Zheng, C., Zheng, Z., Sun, J. et al. (2017). MiR-16-5p Mediates a Positive Feedback Loop in EV71-Induced Apoptosis and Suppresses Virus Replication. *Scientific Reports*, 7, 16422. https://doi.org/10.1038/s41598-017-16616-7

Chapter 2

A proposed artificial intelligence and blockchain technique for solving health insurance challenges

Kuldeep Tomar and Shivani Sharma
NGF College of Engineering &Technology, Palwal, India

2.1 INTRODUCTION

Health insurance is a type of insurance coverage in which the insurance company covers the cost of the insured client's medical and surgical expenses. A contract is signed between the insurer and the client which states that the insurer will cover the cost of the insured client's medical and surgical expenses in return for a particular premium being paid by the client to the insurer. The demand for health insurance is growing rapidly around the world due to the increased rate of illness or diseases and huge medical expenses to cure those diseases.

We currently have a large number of health insurance companies offering various health insurance packages with varying benefits and associated premiums. Still, the health insurance sector is facing difficulties despite a rise in the demand for health insurance and the presence of health insurance companies.

This means there are some problems in the health insurance process that are currently impeding the growth of this sector. The solution of these problems is crucial for the growth of the health insurance sector. To succeed in an increasingly complex environment, organisations must be able to see the big picture while also paying attention to the finer details [1].

2.2 ARTIFICIAL INTELLIGENCE

Artificial intelligence (AI) is created through researching how the human brain works, as well as how people learn, make decisions, and collaborates when attempting to solve a problem. It concentrates on intelligence agents, having the ability to perceive their surrounding plus take actions to maximise their chances of success [2].

A chatbot is a form of software that allows for human interaction and is typically AI-powered. They are also referred to as virtual assistants. A chatbot that responds to questions based on a predetermined scenario is known as a rule-based chatbot. But we also have advanced chatbots which have been taught to converse in a human-like manner using a technique called natural language processing (NLP). Big data, the *Internet of Things (IoT)*, cloud technology, and AI have all had significant impacts on the industry [3].

The insurance industry is frequently thought of as having old procedures, tones of paperwork, and complicated concerns. AI applications can assist businesses in optimizing services and lowering costs, accelerating processes, and making better choices [4]. With the aid of technologies such as AI-enabled chatbots, particularly in the area of digital health insurance, the gap between an insurance provider and their client can be eliminated, creating a direct link between the two. In contrast to traditional insurance, AI has altered how insurers create health insurance policies and enables policyholders to get services more quickly. Insurance companies may streamline their operations, cut expenses, enhance client satisfaction, and boost profitability through the incorporation of AI. Typical time-consuming and laborious procedures such as insuring, claims administration, spam detection, and client services can also be transformed by the use of AI.

2.3 BLOCKCHAIN TECHNOLOGY

Blockchain is defined as " an open, distributed ledger able of recording deals between two parties in an effective, empirical, and endless manner". The data is stored in the form of blocks, with each block pointing to the one before it [5]. The first block is known as the Genesis block, and it doesn't point to the former block, as shown in Figure 2.1. Each Block is made up of a set of transactions and each new block adds to the longest chain.

Blocks are linked together using a hash function which is cryptographically secured in order to form a chain, and this chain serves as a database of historical information accessible to every peer contributing in the network [6], as shown in Figure 2.2.

The public ledger is replicated and distributed among all the nodes present in the network in a synchronized way, assisting network nodes in validating future transactions.

The process of adding new transaction in Blockchain takes place only after the majority of nodes present in the network agree on a specific transaction.

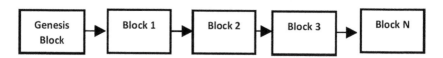

Figure 2.1 Blockchain terminology.

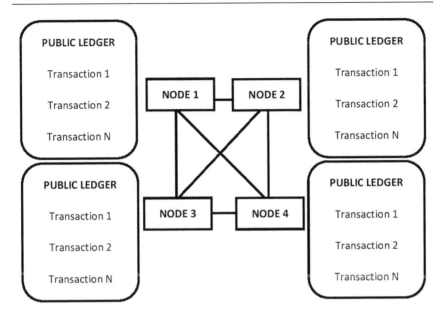

Figure 2.2 Public ledger available to every node.

A special mechanism, known as a Consensus mechanism, is used by a group of nodes on a network to determine which Blockchain transactions are valid and which are not. It is used to achieve a single state of the network among distributed processes [7]. In this way, the public ledger turns as a local copy of global information obtainable to every single peer which gives Blockchain the power to record transactions between two parties efficiently, verifiably and permanently. With its phenomenal scope in the financial and non-financial fields, Blockchain technology is attracting a very diverse audience.

Blockchain is an agreement-oriented, cryptographically secured, distributed public/private ledger technology which stores data in an immutable, irreversible, and resilient manner over a peer-to-peer network [8]. It empowers smart devices to function as autonomous agents capable of performing a variety of transactions [9]. Decentralization, persistence, anonymity, and audibility are all essential attributes of Blockchain technology. With these characteristics, Blockchain can significantly reduce costs and improve efficiency [10]. Permanent record-keeping of the transactions that can be incrementally updated but not erased leaves viewable traces of all chain activities. As a result, the "trust machine" is created by reducing the uncertainty of alternative facts or the truth [11]. Blockchain technology improves data privacy and security by encrypting and distributing data throughout the entire network [12].

Blockchain has been classified in a number of different manners to make it appropriate for use in various fields. Public Blockchain is a completely decentralized, permission-less, open-source system meaning that anyone can access and participate in reading, writing, and auditing the Blockchain. Bitcoin is a peer-to-peer cryptocurrency created by Satoshi Nakamoto in 2009 and based on the public Blockchain concept [13].

Private Blockchain is a permissioned and closed system which is owned by a person or a company. In these circumstances the user requires authentication before entering the network. The read-and-write access are kept permissioned. Users know each other, but they do not trust each other. It has a modular structure that offers high levels of privacy, resiliency, versatility, and expandability. The Linux Foundation has been working on the Hyperledger project since 2015. It is an open-source project which provides a solution to enterprise applications on Blockchain. One of its projects is Hyperledger Fabric, a permissioned Blockchain infrastructure that provides a modular approach which includes smart contract implementation, customizable consensus, and membership services.

This chapter envisions the utilization of AI and Blockchain technology in the health insurance industry. As previously stated, insurance is a form of insurance coverage in which the insurance company pays for the insured client's medical and surgical expenses. At present, the health insurance sector is facing many challenges which are described in this chapter. HealthChain is a new application that uses Blockchain technology to improve security, scalability, and efficiency in the healthcare industry [14]. After analyzing HealthChain as a use case of Blockchain Technology, an idea is being proposed to use Blockchain technology in the health insurance sector. The phases and performance evaluation of the proposed idea using AI and Blockchain technology to deal with the challenges which hinder the growth of the health insurance sector at present are discussed. Then a flow diagram and a comparison table depict how Blockchain technology can provide efficient solutions to problems in the health insurance sector by storing and managing insured clients' records on Blockchain and enhancing its security. There then follows a discussion of how Blockchain is revolutionizing the whole world is discussed.

2.4 THE PRESENT SCENARIO

The health insurance process comprises two separate stages: the Health Insurance Application Process and the Health Insurance Claim Process. The process of buying a health insurance plan and getting insured comes under the Health Insurance Application Process whereas the Health Insurance Claim Process starts when an insured client asks the insurer to cover the cost of the insured client's medical treatment.

The application process begins when the client contacts the health insurance provider to get health insurance. The insurer shows the client various health plans, each with an associated cost estimate. The client selects one plan and fills out the application form, including personal details and medical records. The client may also need to supply any necessary medical test reports. The insurance company then checks the eligibility of the client and makes its response. If the client is deemed to be eligible, they must make their first payment before the plan can be put into action. Once the client has enrolled and made his first payment it takes about 3 weeks for the application form to get processed and for the issuing of the health insurance card.

The claim process starts when the insured client becomes ill and is admitted to the hospital. At this point the insured client needs to show his health insurance card to the Third Party Administrator (TPA) of the hospital. Following that, the hospital needs to submit a pre-authorization application to the insurance company for approval. The insurance company then reviews the insured client's policy to determine whether or not he is eligible for a claim under the policy's terms and conditions. If the insured person is eligible, then the insurance claim will be processed and the insurance company settles the claim.

2.5 CHALLENGES WITH THE EXISTING SYSTEM

The health insurance process has various means of escape which results in the loss of the company or client or both. As the company's website and brochures do not disclose all the terms and conditions of policies, this can lead to a paucity of information among people about health insurance companies and the policies the company offer. Information inadequacy leads to the creation of a gap between people and the company. Clients are easily duped by salespeople leading them to purchase the wrong policies. This may happen when a salesperson is seeking a high commission. This results in a loss of faith on the part of the client and hinders the development of the health insurance sector.

Similarly, a person who wants to have health insurance may provide the wrong medical test report to the health insurer in order to become insured with the widest benefits available. One such example might include a person having diabetes who may take medicine before the test and then show those reports to the health insurer. A person may also fill in an incorrect medical history in the application process to get insured, which results in losses on the part of the company.

It may also be the case that a person has more than one health insurance plan with different companies. Group insurance may get misused by group insurance policyholders as uninsured people may be given treatment at hospitals because the ID cards of several group insurance plans have no photograph for identification.

After getting insured, in order to avail themselves of the benefits of any health insurance plan, some people may become hospitalized for a mild illness because they are insured and will get proper body checkup.

When a health insured client becomes ill and is admitted to a hospital he may have to deposit some amount of cash instead of having cashless health insurance because of the lack of transparency between a health insurance company and hospital which takes time to process the health insurance claim.

If an insured client goes to a hospital for a particular illness, there is a 90% chance that the hospital will impose extra charges [15]. Then higher claims by insured patients lead to increased payouts for the companies, which lead to an increase in premiums. The increase in premiums will place greater pressure on policyholders. The cost of getting new insurance also increases, which affects the growth of the company as a pocket of an individual does not allow them to get insured. According to the IRDAI annual report for 2017–2018, health insurance premiums had risen by more than 20% year on year over the previous three fiscal years [16].

After filling the application form it takes around three weeks for it to get processed, which is a very long time and makes the process of getting insured too slow.

One of the serious points at issue is the long Turn Around Time (TAT). When we discuss cashless treatment, the TAT in the hospital for the payment of the insured client's treatment is 20 days. Sometimes TPAs fail to meet this deadline even after acquiring payment from an insurance company. This could be due to the logistics of dealing with so many hospitals and claims, or it could be because some TPAs do not work on Saturdays, causing claims to be delayed.

It is the foremost responsibility of health insurance companies to protect the data of their clients from data breaches. The cost of cybercrime quadrupled between 2013 and 2015, but a significant proportion of cyberattacks goes unnoticed. Global cybercrime expenses are predicted to increase by 15% yearly, reaching 10.5 trillion dollars in annual costs by 2025, according to Cybersecurity Ventures, a leading cybersecurity economy research firm [9]. Data confidentiality and data authentication must not be compromised. The HIPAA (Health Insurance Portability and Accountability Act of 1996) privacy rule standardizes the usage and disclosure of Protected Health Information (PHI) held by covered entities (generally health insurers, medical service providers that engage in certain transactions) [17]. Data breaches lead to violation of HIPAA rules. In accordance with the U.S. Department of Health and Human Service Office for Civil Rights, various health insurance companies fall prey to data breaches, which included, among others, the Anthem Blue Cross Health Insurance Company in 2015, which affected 78.8 million customers, and the Premera Blue Cross in 2015, which affected more than 11 million [18]. A further data breach at HealthCare.gov, which affected approximately 75,000 people, exposed personal details, along with health coverage details [19].

The insurance regulatory and development authority (IRDAI) is responsible for regulating, promoting, and ensuring the organized expansion of the

Table 2.1 Sector-wise net incurred claims ratio under health insurance [16]

Sector	2013–2014	2014–2015	2015–2016	2016–2017	2017–2018
Public sector General Insurance	106%	112%	117%	122%	108%
Private sector General Insurance	87%	84%	81%	84%	80%
Stand along Health Insurance	67%	62%	58%	68%	62%
Industry Average	97%	101%	102%	106%	94%

https://www.irdai.gov.in/ADMINCMS/cms/frmGeneral_NoYearList.aspx?DF=AR&mid=11.1

industry. According to IRDAI, the present high claim ratio is a serious issue [20]. There has been a high Incurred Claim Ratio (ICR) in the past few years, which is above 90% according to IRDAI, making the health insurance sector less profitable. The net ICR was 97% in 2013–2014, 101% in 2014–2015, 102% in 2015–2016, 106% in 2016–2017 and 94% in 2017–2018 [16]. Table 2.1 shows the net incurred ratio by sector for health insurance.

Analysis of this data show that challenges in the health insurance process are impacting the health insurance sector to a larger extent and must be addressed. Some of these problems lead to an increase in ICR, which makes the health insurance sector unprofitable and people starts to lose interest in this sector which hinders the growth of this sector. Other problems result in a degradation in the quality of service which the company provides to the customer; this indirectly affects the health insurance sector as a customer starts to lose faith in the company.

2.6 THE SOLUTION: ARTIFICIAL INTELLIGENCE AND BLOCKCHAIN TECHNOLOGY

"The Blockchain is an incorruptible digital ledger of economic transactions that can be programmed to record not just financial transactions but virtually everything of value" according to Don and Alex Tapscott, the authors of *Blockchain Revolution* (2016). We have discussed various drawbacks of the health insurance sector and its process. Now we must understand how Artificial Intelligence and Blockchain technology can be utilized to address the challenges that the health insurance industry is currently experiencing.

The main characteristics of AI distinguish it from competing technologies and allow it to excel in every industry. If one or more of the features listed below satisfy the use case's requirements, you should apply artificial intelligence in that field or use case.

- Automation: AI enables machines to carry out tasks that would normally be performed by people. AI redefines automation as it tirelessly completes essential automated tasks.
- Smart Decision-making: Artificial Intelligence helps in smart decision-making by improving the effectiveness and quality of decision-making.
- Enhanced customer experience as being able to gather and process client information in real-time aids in better understanding consumer behaviour and demands, which in turn aids in developing a plan for a tailored customer experience.

Blockchain technology has some key features which make it an astonishing technology proving itself in every field and distinguishable it from others. You should use Blockchain technology in any field or use case if one or more feature presented below matches with your requirements:

- Transparency: When you want data to be available to everyone so that every node in the network can verify the data. Every node has equal control over the network. No node is superior to any other and there are no chances of a discrepancy.
- Decentralization: The network is designed in such a manner that there is no overall authority governing the network. Rather, you choose to hand over the data management to the network because you have a lack of trust in the parties involved.
- Tamper with evidence: A time-stamp signifies that the parts of any data file which existed at that moment really have not updated. A different person edits the document from time to time. Thus, we time-stamp the digital signature to retain a record of who has edited the document and at what time.

So, AI and Blockchain has the potential to solve the issues of the health insurance sector by making it automated, transparent, decentralized, and tamper-proof.

2.7 PROPOSED IDEA

After analysing the issues and identifying them as challenges to the development of the health insurance sector, it can be said that we need better control over the whole process of getting insured to getting a claim. We need processes to be more efficient and effective in nature, improving the overall experience of the customer by automating the tasks. A solid framework for preventing security breaches is required. More transparent, secure, and efficient processes are desired. We need a strong identification mechanism so that the frauds could be minimized and we must be focused on betterment in service to customers. Effective regulations and rules need to be established and followed for the benefit of the company and customers.

AI is a technology which is making huge strides in every sector. Thus, the chatbot which contains an AI algorithm can perform a client/user interaction via text or voice interface. It is also known that an AI-enabled chatbot can be trained easily and provides facilities of live chat or information flow. It has the capability to provide numerous advantages to the entire health insurance process. Blockchain has recently emerged and is operating as a pioneer of the digital age by bringing a fresh point of view to system security, resilience and effectiveness. HealthChain is among the new applications and it uses Blockchain technology to improve safety, expandability and effectiveness in the healthcare industry. The HIPAA privacy rule establishes standards for ensuring the confidentiality of individual citizens' PHI and offers patients access to that information. Despite the conventional encryption and passcode configuration adhered by HIPAA for a covered entity, breaches may still take place and PHI may be compromised. The importance of the correct use and incorporation of new devices with access to PHI is paramount in a productive healthcare era. Blockchain technology has the potential to increase system performance while also ensuring the security and expandability. HealthChain, which is power-driven by Blockchain, benefits greatly from the modular architecture of Hyperledger fabric which supports confidentiality, scalability, and security in health informatics. Its use of smart contracts ensures proper permission and privilege setting on its permissioned network [14].

After analyzing the HealthChain use case of Blockchain technology, it is proposed to use Blockchain technology in the health insurance sector to deal with the challenges which hinder the growth of the present-day health insurance sector.

In today's scenario, a health insurance company holds the data of each insured client in his company's database. If an insured client wants to obtain some information regarding his policy he needs to contact the health insurance company. If an uninsured client wants to get information regarding policies he needs to contact the health insurance company. In addition, the hospital needs to contact the health insurance company during the claiming process. Thus, everyone depends on the health insurance company to provide details, as we envisage in Figure 2.3.

According to the proposed idea, the health insurance company stores the data of each insured client on Blockchain and automation is provided through AI. As the data is stored over Blockchain, this works as a public ledger. If an insured client wants to get some information regarding his policy he may go to Blockchain. If the hospital needs some information for the claiming process, the hospital may go to Blockchain. So we do not depend on the health insurance company to provide details regarding the policy.

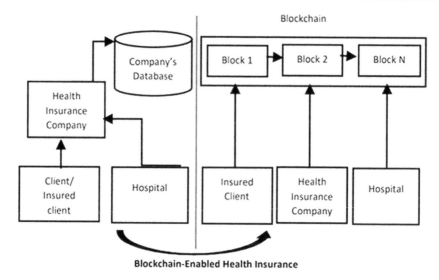

Figure 2.3 Blockchain-enabled health insurance.

Instead, the insured client, the insurer, and the hospital may go to Blockchain to check for any detail regarding the insured client. Also, AI can help clients to get the necessary information when required.

2.8 PROPOSED IDEA PHASES AND PERFORMANCE ANALYSIS

2.8.1 Phase I: shop and estimate

A client may contact a health insurance salesperson (insurer) or a chatbot-enabled website if he wants to become insured. Both places will act as a source of information for the client. The salesperson will show the client various health plans and will provide the client with an estimate of the costs involved. Alternatively, the client may go to the website where he can communicate with the chatbot to fetch all the information regarding all plans with clear-cut terms and conditions and estimates. Also, the client can fill his details on the website with a selection of one policy. The chatbot can then decide whether or not the client is eligible for that particular policy. The entire data will be kept in Blockchain to improve the safety, resiliency, and efficiency of the system.

Performance analysis of Phase I:

- Salesperson cannot mislead clients or sell incorrect products in order to avoid pressure from their higher authority and receive incentives. A bad product is sometimes sold for a higher commission.

- The chatbot helps clients to buy the right policy according to the client's requirement.
- Sufficient information regarding policies creates awareness among the client.
- Whenever the client accesses the website and fills in details a database will be created which will help companies to analyze the interests of the client and then plan policies accordingly, which will help in the growth of the health insurance sector.

2.8.2 Phase II: apply and enroll

After choosing the plan the client needs to apply for the desired plan by filling an application form which includes personal details and medical records. Personal details include identity proof for the client, the iris scan of a client, the thumb print of a client, and other basic details. After that, the client goes for necessary health checkups. The client needs to get the test done at the health insurance pathology center. Then the whole data is fed to the Blockchain-enabled platform which checks whether or not the client is eligible for a specific health insurance plan and responds accordingly. If the client is eligible and his application has been accepted, he gets informed, and if not then also.

Performance analysis of Phase II:

- A client who wants to have health insurance cannot provide incorrect medical test reports to the health insurer as tests are conducted at their center.
- A client cannot fill in wrong medical history in the application process to get insured as everything is getting stored on Blockchain-enabled platform, if he does so, he may face legal issues in the future (at the time of claim).
- If a person already has one health insurance, we will have proper record as everything is stored on Blockchain which acts as a public ledger.
- Advanced identification mechanisms like an iris scan or a thumbprint will be used for the proper authentication of the insured clients at the time of any claim. Group policy will not be misused.

2.8.3 Phase III: make your first payment

If the client is eligible for the plan he chooses, he needs to make his first payment before his health insurance begins. Everything is stored on Blockchain. The payment will also act as a transaction and stored on Blockchain.

Performance analysis of Phase III:

- It results in transparent processing and time-efficient solutions. We do not need to wait for weeks to get the process completed.

2.8.4 Phase IV: cards and coverage

Once everything is undertaken the client will be allotted his health insurance card with sufficient information recorded on it. Now the client becomes an insured client and can access his information on his Blockchain block using keys (password) allotted to him.

Performance analysis of Phase IV:

- This results in very rapid operation. In such circumstances, there is no need to wait for three weeks for the application to be processed.
- An insured person has access to see all necessary details on his Blockchain-enabled block, which increases the level of transparency.

2.8.5 Phase V: claim process

The claim process starts when an insured client goes to the hospital after getting a medical emergency. As the data of the patient (the insured client) is stored over Blockchain the hospital will be able to check the eligibility of patients (insured clients) getting a claim in no time. The hospital will be able to see the insured client's detail on Blockchain. A smart contract will be executed at the Blockchain platform and the hospital will get to know whether or not the insured client is insured for that particular medical emergency. This creates greater transparency. After the execution of a smart contract, if the patient (insured client) is eligible for a claim, the hospital will provide cashless treatment to patients (insured client) and the fastest treatment possible.

Each request to Blockchain will be treated as a transaction and each record of a patient (insured client) will be stored on the system. This gives the benefit that no hospital would ever try to fill in wrong information for the patient (insured client) as everything is being stored on Blockchain. If a hospital does so, it might be subject to heavy penalties.

In addition, in order to enhance customer experience or to improve the level of customer assistance, AI-powered chatbots that can imitate human conversations will be employed.

To better understand consumer inquiries and respond to various questions regarding insurance claims or product selection, chatbots will be deployed employing sentiment analysis and natural language processing.

Performance analysis of Phase V:

- No hospital would ever try to charge you more as you are insured.
- No hospital will try to store the wrong information on Blockchain.
- There will be an improvement in the level of service to patients (insured client).

2.9 FLOW DIAGRAM OF PROPOSED IDEA

After discussing these problems we propose a solution named "Blockchain and Artificial Intelligence enabled Health Insurance". Blockchain provides a more secure, transparent, tamper-proof system where we can store our health insurance records. Being immutable, it would be impossible to tamper with the data stored there. There would be better control over the whole process of getting insured to getting a claim. Our idea is to use permission Blockchain technology in the health insurance process. We have a closed system in permission Blockchain which is possessed by a person or company. The user needs to be authenticated before entering the network. The read and write access are kept permissioned. Users know each other, but do not trust each other. It has a modular structure that offers a high degree of privacy, resiliency, versatility, and expandability. The flow diagram of Blockchain-enabled health insurance is shown in Figure 2.4, which illustrates the read/write access of each node in the network.

This concept could be implemented using Hyperledger Fabric, a permissioned Blockchain infrastructure developed by IBM. Membership service provides identities within the distributed Fabric network. A client may access a Blockchain-enabled health insurance website and can apply through a health insurance salesperson. The health insurance salesperson, insured

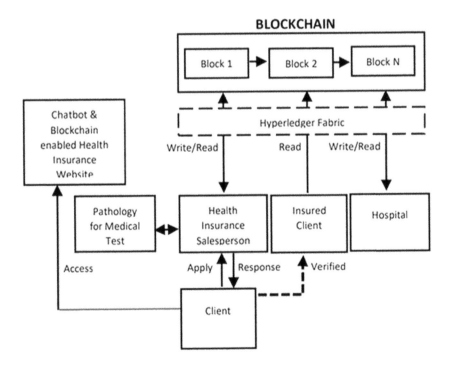

Figure 2.4 Flow diagram of proposed idea.

client and hospital are all identified, and each of them have different levels of access to Blockchain.

2.10 RESULT ANALYSIS

To create a prototype, the coding has been done in java. We computed the hash of the entire block using the SHA-256 hash algorithm.

The process begins with the creation of the Genesis Block, which is the first block in the Blockchain. The Genesis Block contains elementary parameters such as the level of difficulty, the consensus algorithm, etc. which are used to mine future blocks. The Genesis Block is the only block in the Blockchain which does not refer to the previous block simply because it does not have any previous block.

The Genesis Block for our system was created on 1 April 2019. The Genesis Block hash is calculated as 111273125, as shown in Figure 2.5.

After the creation of the genesis block, a new block is inserted to the existing Blockchain storing data of the client, Shivani Sharma, including all of her policy details: the client's name, policy type, date of the policy, age, level of insurance, validity, coverage. A previous hash of Blockchain is used to compute the present hash of the block. The previous block hash is 111273125 and the current block hash is -1996524128, as shown in Figure 2.6.

Some new block is also inserted to the existing Blockchain, storing data of different clients with their policy details, which again includes the name of the client, policy type, date of a policy, age, amount of getting insured, validity, coverage, etc. The following, Figures 2.7 and 2.8, shows a health insurance Blockchain with the records of six clients.

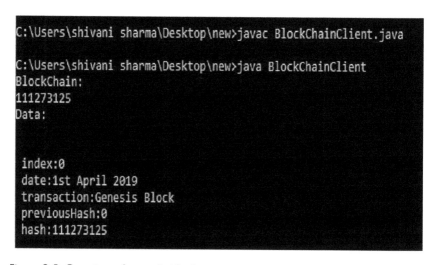

Figure 2.5 Creation of genesis block.

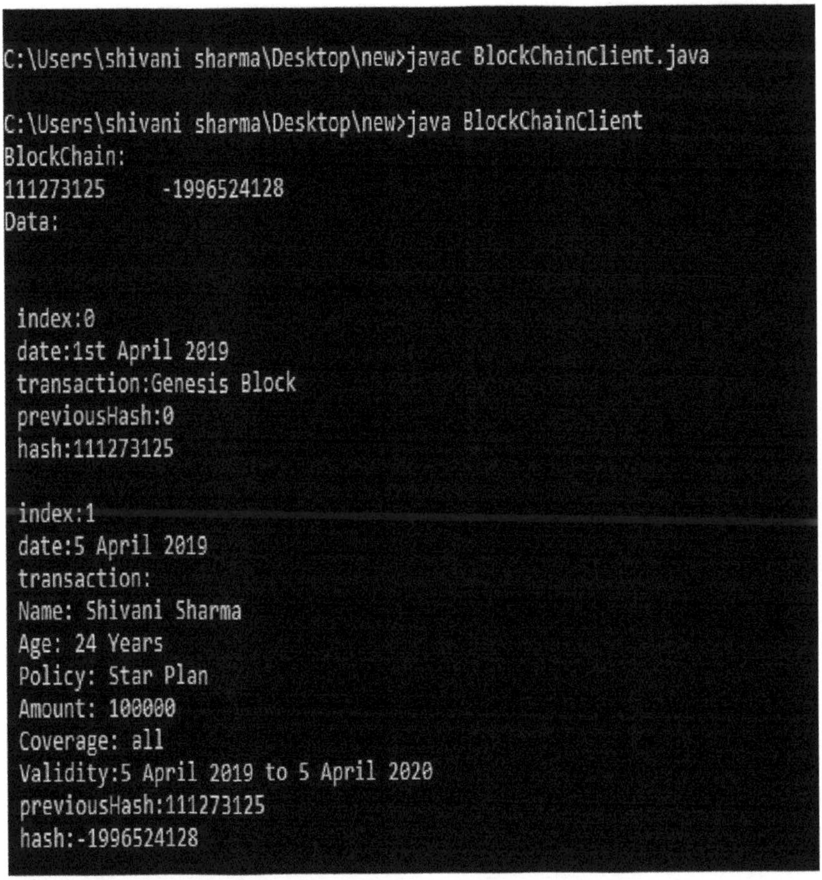

```
C:\Users\shivani sharma\Desktop\new>javac BlockChainClient.java

C:\Users\shivani sharma\Desktop\new>java BlockChainClient
BlockChain:
111273125      -1996524128
Data:


 index:0
 date:1st April 2019
 transaction:Genesis Block
 previousHash:0
 hash:111273125

 index:1
 date:5 April 2019
 transaction:
 Name: Shivani Sharma
 Age: 24 Years
 Policy: Star Plan
 Amount: 100000
 Coverage: all
 Validity:5 April 2019 to 5 April 2020
 previousHash:111273125
 hash:-1996524128
```

Figure 2.6 Creation of 2nd block.

Now we tried to check whether or not the whole health insurance Blockchain is valid. If anyone attempts to modify any data in any block, then the Blockchain becomes tampered with. While adding the 8th block in the Blockchain we checked whether or not the Blockchain is authentic. If the Blockchain is tampered with by someone by changing any data in the block it returns false Blockchain, otherwise it is true Blockchain. At present, the Blockchain has not been tampered with, as shown in Figures 2.9 and 2.10.

After the creation of the 8th block, an intruder tried to falsify the data of client Shivani Sharma in the 2nd block, which resulted in the loss of authenticity of Blockchain and the whole Blockchain becoming invalid. Blockchain uses the concept of a Merkle tree. If we try to change any information in any block this results in the creation of a different block hash. The previous hash of block is used to create a new block. When an intruder attempts to alter the data of any block this leads to the generation of a different hash of the block which invalidates the whole successor chain, as shown in Figure 2.11.

```
index:0
date:1st April 2019
transaction:Genesis Block
previousHash:0
hash:111273125

index:1
date:5th April 2019
transaction:
Name: Shivani Sharma
Age: 24 Years
Policy: Star Plan
Amount: 100000
Coverage: all
Validity:5 April 2019 to 5 April 2020
previousHash:111273125
hash:-1996524128

index:2
date:6th April 2019
transaction:
Name: Kuldeep Tomar
Age: 34 Years
Policy: Galaxy Plan
Amount: 1000000
Coverage: all
Validity:6 April 2019 to 6 April 2020
previousHash:-1996524128
hash:513986884

index:3
date:8th April 2019
transaction:
Name: Pranit Sharma
Age: 20 Years
Policy: Floater Plan
Amount: 10000
Coverage: Specific (Cancer and Tumuor after 6 months)
Validity:8 April 2019 to 8 April 2020
previousHash:513986884
hash:14072546
```

Figure 2.7 Creation of 4th block.

After tampering with the data, all the successor blocks becomes invalid and the resultant Blockchain is shown below in Figure 2.11.

The output Blockchain, after tampering with data, resulted in the invalidation of successor Blockchain from the block which was tampered with. From this, we conclude that it is unrealistic to intrude in the case of data stored on Blockchain. This immutability feature of Blockchain makes it a robust technology in providing prevention against security breaches. The public ledger feature makes the processes more transparent, secure, and efficient.

```
index:4
date:9th April 2019
transaction:
Name: Priya Kaushik
Age: 25 Years
Policy: Star Plan
Amount: 100000
Coverage: all
Validity:9 April 2019 to 9 April 2020
previousHash:14072546
hash:-1496264871

index:5
date:10th April 2019
transaction:
Name: Shobha Shukla
Age: 30 Years
Policy: Floater Plan
Amount: 10000
Coverage: Specific (Cancer and Tumuor after 6 months)
Validity:10 April 2019 to 10 April 2020
previousHash:-1496264871
hash:1467151331

index:6
date:11th April 2019
transaction:
Name: Pooja Sharma
Age: 24 Years
Policy: Star Plan
Amount: 100000
Coverage: all
Validity:11 April 2019 to 11 April 2020
previousHash:1467151331
hash:158556213
```

Figure 2.8 Creation of 7th block.

All the issues which the health insurance sector is facing at present can be efficiently and effectively solved by using Blockchain technology.

2.11 FLOW DIAGRAM OF OUTPUT

The entire Blockchain with a record of each insured client has been shown in Figures 2.12, 2.13, and 2.14. The first block is known as the genesis block. The hash of the 1st block (genesis block) is used to compute the hash of the 2nd block. The hash of the 2nd block is used to compute the hash of the 3rd block and so on.

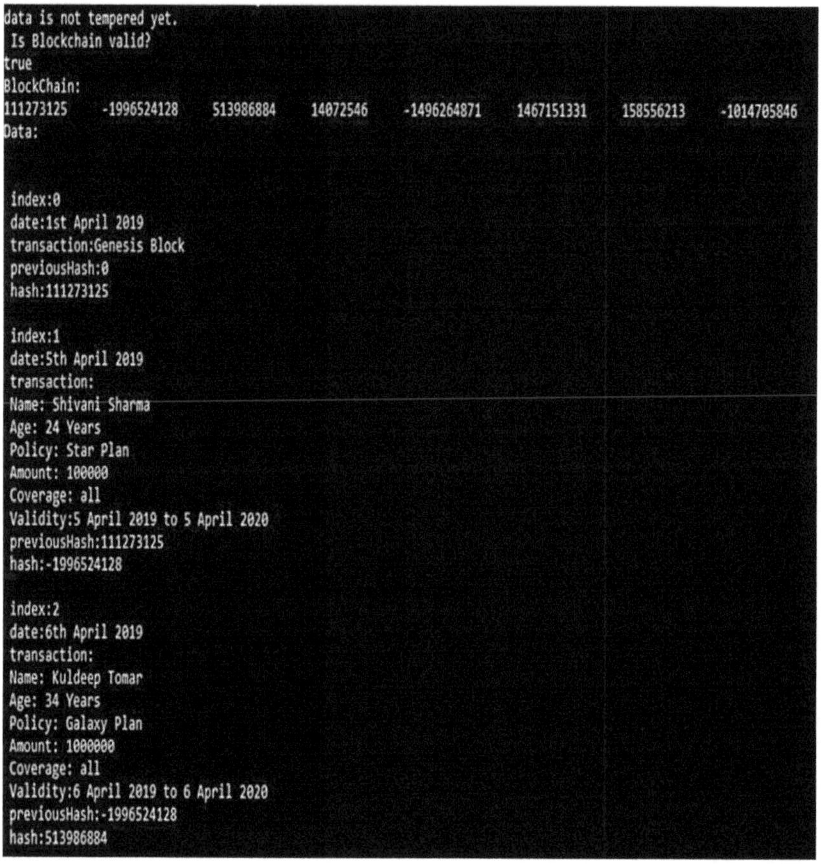

```
data is not tempered yet.
Is Blockchain valid?
true
BlockChain:
111273125    -1996524128    513986884    14072546    -1496264871    1467151331    158556213    -1014705846
Data:

index:0
date:1st April 2019
transaction:Genesis Block
previousHash:0
hash:111273125

index:1
date:5th April 2019
transaction:
Name: Shivani Sharma
Age: 24 Years
Policy: Star Plan
Amount: 100000
Coverage: all
Validity:5 April 2019 to 5 April 2020
previousHash:111273125
hash:-1996524128

index:2
date:6th April 2019
transaction:
Name: Kuldeep Tomar
Age: 34 Years
Policy: Galaxy Plan
Amount: 1000000
Coverage: all
Validity:6 April 2019 to 6 April 2020
previousHash:-1996524128
hash:513986884
```

Figure 2.9 Checking the validity of Blockchain till 3rd block.

When the data of client Shivani Sharma was tampered with in the 2nd block of the blockchain, it leads to a change in the hash of the 2nd block. This is because the previous block's hash is used to compute the hash of the current block. In our example the entire Blockchain was invalidated after the 2nd block. The resultant Blockchain after tampering is shown in Figures 2.15, 2.16 and 2.17.

2.12 COMPARISON TABLE

After analyzing the performance of each phase in the proposed idea, we can say that Blockchain technology can provide an efficient and effective solution to problems of the health insurance sector. Table 2.2 compares the present system to the proposed model in the health insurance sector.

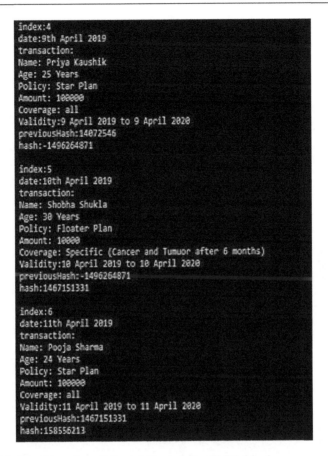

```
index:4
date:9th April 2019
transaction:
Name: Priya Kaushik
Age: 25 Years
Policy: Star Plan
Amount: 100000
Coverage: all
Validity:9 April 2019 to 9 April 2020
previousHash:14072546
hash:-1496264871

index:5
date:10th April 2019
transaction:
Name: Shobha Shukla
Age: 30 Years
Policy: Floater Plan
Amount: 10000
Coverage: Specific (Cancer and Tumuor after 6 months)
Validity:10 April 2019 to 10 April 2020
previousHash:-1496264871
hash:1467151331

index:6
date:11th April 2019
transaction:
Name: Pooja Sharma
Age: 24 Years
Policy: Star Plan
Amount: 100000
Coverage: all
Validity:11 April 2019 to 11 April 2020
previousHash:1467151331
hash:158556213
```

Figure 2.10 Checking the validity of Blockchain till 7th block.

```
index:7
date:1st march 2019
transaction:
Name: Rohit Sharma
Policy: Basic  Plan
Amount: 50000
Coverage: limited
previousHash:158556213
hash:-1014705846
```

Figure 2.11 Creation of 8th block.

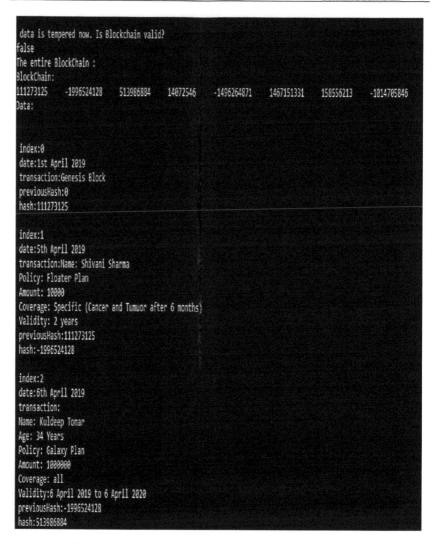

Figure 2.12 Blockchain after tampering of data.

Table 2.2 shows that Blockchain provides a more secure, transparent, immutable system where we can store our health insurance records. It provides modular architecture delivering high degrees of confidentiality, resiliency, flexibility, and scalability.

```
current block hash:-1527604796
Block mined with hash :-1527604796
true
BlockChain:
111273125      -1527604796
Data:


 index:0
 date:1st April 2019
 transaction:Genesis Block
 previousHash:0
 hash:111273125

 index:1
 date:1st feb 2017
 transaction:Name: Shivani Sharma
 Policy: Floater Plan
 Amount: 10000
 Coverage: Specific (Cancer and Tumuor after 6 months)
 Validity: 2 years
 previousHash:111273125
 hash:-1527604796
```

Figure 2.13 The output Blockchain after tampering of data.

2.13 CONCLUSION

Blockchain technology and Artificial Intelligence are a revolutionary technology which is growing rapidly in many different fields. Artificial Intelligence is allowing machines to work in a powerful way leading to the automation of tasks and improved efficiency. Blockchain has established new standards in information security and founded itself as a powerful tool that can be used to combat security issues. This technology had begun its journey with bitcoin and has then moved into different directions to provide feasible solutions to every problem. We have proposed the use of Artificial Intelligence and Blockchain technology in the health insurance sector. This has led to the adoption of Artificial Intelligence and Blockchain technology in the health insurance sector which enabled us to fight challenges which are hindering the growth of the health insurance sector.

Index: 0
Date: 1ˢᵗ April 2019
Transaction:
Genesis Block
Previous Hash: 0
Current Hash: 111273125

Index: 1
Date: 5ᵗʰ April 2019
Transaction:
Name: Shivani Sharma
Age: 24 Years
Policy: Star Plan
Amount: 100000
Coverage: All
Validity: 5ᵗʰ April 2019 to 5ᵗʰ April 2020
Previous Hash: 111273125
Current Hash: -1996524128

Index: 2
Date: 6ᵗʰ April 2019
Transaction:
Name: Kuldeep Tomar
Age: 34 Years
Policy: Galaxy Plan
Amount: 1000000
Coverage: All
Validity: 6ᵗʰ April 2019 to 6ᵗʰ April 2020
Previous Hash: -1996524128
Current Hash: 513986884

Figure 2.14 Health insurance Blockchain till 3rd block.

Index: 3
Date: 8th April 2019
Transaction:
Name: Pranit Sharma
Age: 20 Years
Policy: Floater Plan
Amount: 10000
Coverage: Specific (Cancer and tumor
after 6 months)
Validity: 8th April 2019 to 8th April 2020
Previous Hash: 513986884
Current Hash: 14072546

Index: 4
Date: 9th April 2019
Transaction:
Name: Priya Kaushik
Age: 25 Years
Policy: Star Plan
Amount: 100000
Coverage: All
Validity: 9th April 2019 to 9th April 2020
Previous Hash: 14072546
Current Hash: -1496264871

Index: 5
Date: 10th April 2019
Transaction:
Name: Shobha Shukla
Age: 30 Years
Policy: Floater Plan
Amount: 10000
Coverage: Specific (Cancer and tumor
after 6 months)
Validity: 10th April 2019 to 10th April
2020
Previous Hash: -1496264871
Current Hash: 1467151331

Figure 2.15 Health insurance Blockchain till 6th block.

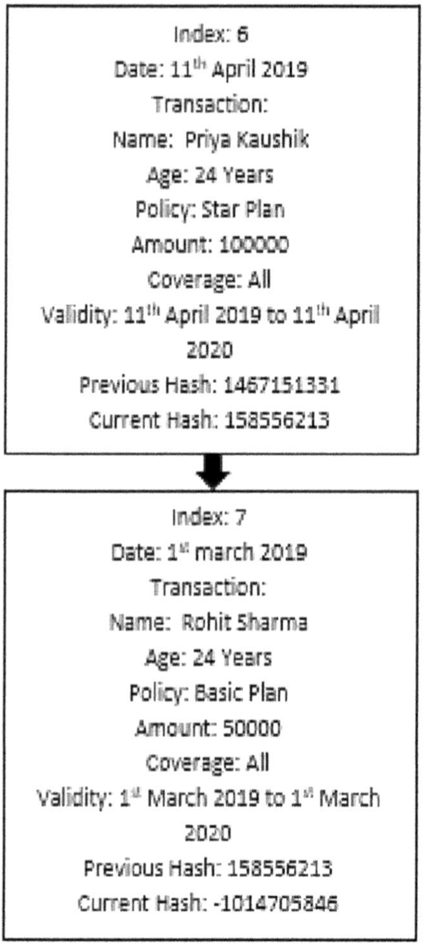

Index: 6
Date: 11th April 2019
Transaction:
Name: Priya Kaushik
Age: 24 Years
Policy: Star Plan
Amount: 100000
Coverage: All
Validity: 11th April 2019 to 11th April
2020
Previous Hash: 1467151331
Current Hash: 158556213

Index: 7
Date: 1st march 2019
Transaction:
Name: Rohit Sharma
Age: 24 Years
Policy: Basic Plan
Amount: 50000
Coverage: All
Validity: 1st March 2019 to 1st March
2020
Previous Hash: 158556213
Current Hash: -1014705846

Figure 2.16 Health insurance Blockchain till 8th block.

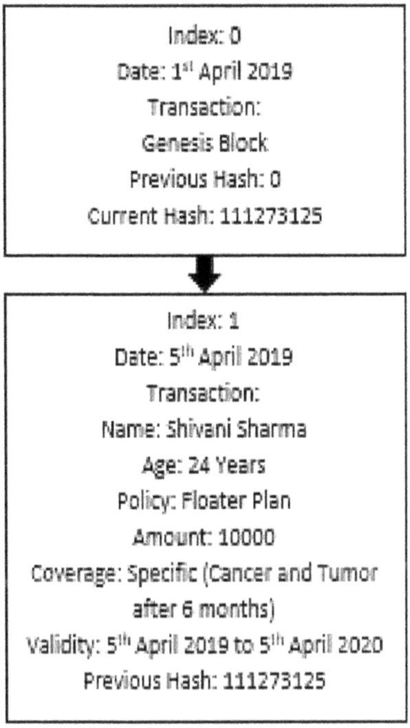

```
Index: 0
Date: 1st April 2019
Transaction:
Genesis Block
Previous Hash: 0
Current Hash: 111273125
```

```
Index: 1
Date: 5th April 2019
Transaction:
Name: Shivani Sharma
Age: 24 Years
Policy: Floater Plan
Amount: 10000
Coverage: Specific (Cancer and Tumor
after 6 months)
Validity: 5th April 2019 to 5th April 2020
Previous Hash: 111273125
```

Figure 2.17 Health insurance Blockchain after the tamper of data.

Table 2.2 Comparison of the current and proposed systems

Basis of comparison	Current system	Proposed system
Immutable	Data can be changed after it has been saved in the company's database because data is not permanent.	Once data is stored in Blockchain, it cannot be tampered with because data is permanent.
Efficiency	At present health insurance process includes manual processing which consumes a lot of time making the system less efficient.	Blockchain provides a fast solution with the help of smart contracts.
Fault Tolerance	Being data stored in a company's database, there is a risk of centralized failure	There is no single point of failure because the system is distributed.
Response Time	Most of the Traditional mechanisms are time consuming and thus client satisfaction is at stake.	Using AI enabled chatbot, typical time-consuming and laborious procedures like insuring, claims administration, spam detection, and client service can also be transformed.

(Continued)

Table 2.2 (Continued)

Basis of comparison	Current system	Proposed system
Transparency	Data is not available to each node which makes the system less transparent.	Each node has a public ledger making system more transparent and gives each node the power to verify each and every transaction.
Irreversible and Auditable	Data once entered can be altered or undone.	Data once entered cannot be altered or undone.
Security	The present system is less secure and prone to attacks.	Cryptography secures data by providing privacy protection, truthfulness, identity verification, and non-repudiation.
Non-Repudiation	There is no way to trace someone's activity. So anyone can deny about validity of something.	Everything is recorded on Blockchain so someone can't deny about the validity of something.
Time-stamped Documents	As there is no rule for time stamping, there may be chances of forgery.	Time stamping Records protect the system from falsification, and time stamping the entries reduces scam.

REFERENCES

[1] Vimlesh Kumar Ojha, Sanjeev Goyal, and Mahesh Chand, "Study on Data-Driven Decision-Making in Entrepreneurship", *Principles of Entrepreneurship in the Industry 4.0 Era.* Edition:1st Edition, eBook ISBN: 9781003256663, First Published 2022 CRC Press, pp. 75–88. https://doi.org/10.1201/9781003256663. https://www.taylorfrancis.com/books/edit/10.1201/9781003256663/principles-entrepreneurship-industry-4-0-era-rajender-kumar-rahul-sindhwani-tavishi-tewary-paulo-davim

[2] Jahanzaib Shabbir, and Tarique Anwer, "Artificial Intelligence and Its Role in Near Future", *Journal of Latex Class Files*, Vol. 14, No. 8, August 2015.

[3] Vimlesh Kumar Ojha, Sanjeev Goyal, and Mahesh Chand, "A Review of Data-Driven Decision Making in Advance Manufacturing Systems", *Computer Integrated Manufacturing Systems*, Vol. 28, No. 11, pp. 172–182, 2022.

[4] Steffen Hehner, and Boris Körs, "Artificial intelligence in health insurance: Smart claims management with self-learning software", September 1, 2017, McKinsey & Company. https://www.mckinsey.com/industries/healthcare/our-insights/artificial-intelligence-in-health-insurance-smart-claims-management-with-self-learning-software

[5] www.nptel.ac.in

[6] Nofer Michael, Gomber Peter, Hinz Oliver, and Schiereck Dirk, "Blockchain", DOI: 10.1007/s12599-017-0467-3, 2017 Springer Fachmedien Wiesbaden.

[7] Halpin Harry, and Piekarska Marta, "Introduction to Security and Privacy on the Blockchain", *2017 IEEE Europeon Symposium on Security and Privacy Workshops (EuroS&PW)*, DOI: 10.1109/EuroSP.2017.26.43, 2017 IEEE.

[8] Chatterjee Rishav, and Chatterjee Rajdeep, " An Overview of the Emerging Technology: Blockchain", *International Conference on Computational Intelligence and Networks*, DOI: 10.1109/CINE.2017.33, 2017 IEEE.

[9] Singh Sachchidanand, and Singh Nirmala, "Blockchain: Future of Financial and Cyber Security", *2nd International Conference on Contemporary Computing and Informatics (IC3I)*, 978-1-5090-5256-1/16/, 2016 IEEE.

[10] Zheng Zibin, and Xie Shaoan, "An Overview of Blockchain Technology: Architecture, Consensus, and Future Trends", *6th International Congress on Big Data*, 978-1-5386-1996-4/17, DOI: 10.1109/BigDataCongress.2017.85, 2017 IEEE.

[11] Beck Roman, "Beyond Bitcoin: The Rise of Blockchain World", *IEEE COMPUTER SOCIETY*, 0018-9162/18/$33.00 ©, 2018 IEEE.

[12] Elisa Noe, and Yang Longzhi, "A Framework of Blockchain-Based Secure and Privacy-Preserving E-government System", *Wireless Networks*, DOI: 10.1007/s11276-018-1883-0, Springer.

[13] Satoshi Nakamoto, "Bitcoin: A Peer-to-Peer Electronic Cash System". https://bitcoin.org/bitcoin.pdf

[14] Tareq Ahram, Arman Sargolzaei, Saman Sargolzaei, Jeff Daniels, and Ben Amaba, "Blockchain Technology Innovations", *2017 IEEE Technology & Engineering Management Conference (TEMSCON)*, 978-1-5090-1114-8/17/$31.00 ©2017 IEEE, DOI: 10.1109/TEMSCON.2017.7998367. https://www.researchgate.net/publication/318894127_Blockchain_technology_innovations

[15] Suman Devi, and Vazir Singh Nehra, "The Problems with Health Insurance Sector in India", *Paripex – Indian Journal Of Research*, Vol. 4, No. 3, March 2015.

[16] https://www.irdai.gov.in/ADMINCMS/cms/frmGeneral_NoYearList.aspx?DF=AR&mid=11.1

[17] https://en.wikipedia.org/wiki/Health_Insurance_Portability_and_Accountability_Act

[18] https://digitalguardian.com/blog/top-10-biggest-healthcare-data-breaches-all-time

[19] https://www.cnbc.com/2018/11/09/healthcaregov-data-breach-exposed-personal-details-of-75000.html

[20] Ramaiah Itumalla, G. V. R. K. Acharyulu, and L. Kalyan Vishwanath Reddy, "Health Insurance in India: Issues and Challenges", *International Journel of Current Research*, Vol 8, No. 02, pp. 26815–26817, February, 2016.

Chapter 3

Logistics performance measurement

A data envelopment analysis using the logistics performance index 2018 data

M. Mujiya Ulkhaq

Diponegoro University, Kota Semarangi, Indonesia

3.1 INTRODUCTION

Logistics performance concerns how efficiently supply chains connect firms to domestic and international opportunities (Arvis et al., 2018). The logistics industry is one of the major development pillars recognized by policymakers around the world. Inefficient logistics drives up the costs of trade and limits the potential for global integration. For emerging nations attempting to participate in the global market, this is a significant burden.

The logistics performance index (LPI) published by the World Bank is considered a benchmarking tool to assist countries find the opportunities as well as the challenges they face in their logistics performance and what they can do to improve it. The objective is to assess the efficiency of logistics according to the survey of logistics professionals. When the LPI was initially released in 2007, it sparked a discussion about the role that logistics plays in the expansion of the global economy. This index is developed by including the six following indicators: infrastructure, customs, logistics service quality and competence, international shipments, timeliness, as well as tracking and tracing.

Logistics efficiency is assessed by implementing data envelopment analysis (DEA). In this study the recent LPI 2018 data is used. In this context the term efficiency refers to the ratio of output to input (Cooper et al., 2006) and it is frequently used as a measure of performance evaluation. The efficiency is mainly measured by the frontier methods in two forms: parametric, e.g., stochastic frontier analysis (SFA); and non-parametric, e.g., data envelopment analysis (DEA). The frontier models have attracted significant attention from researchers since the frontier concept demonstrates the crucial characteristics of measuring efficiency, i.e., it attempts to measure how well an organization produces maximum output by consuming minimum inputs (De Witte & López-Torres, 2017). The SFA has been long criticized for relying on restrictive assumptions about the distribution of the error term and the functional form. On the other hand, the DEA relaxes those assumptions; in addition, it can handle multiple outputs more easily than the SFA. For this reason, the DEA is used to assess country's efficiency as the logistics

DOI: 10.1201/9781003462163-3

performance measurement. As this study allows for international comparisons, it is expected as a means of benchmarking, comparing logistics performance against a set of counterparts in different countries. In addition, this
study is expected to show that the decision-making process can be driven by
such a big data set (Ojha et al., 2022, 2023).

DEA has been widely used to assess efficiency in several sectors, such as
education (Agasisti & Zoido, 2018, 2019; Santín & Sicilia, 2018), industrial
sectors (Handayani et al., 2020; Ulkhaq, 2022; Ulkhaq & Pratiwi, 2022),
small and medium-sized enterprises (Pramono et al., 2019; Sari et al., 2018),
agriculture (Moradi et al., 2018; Toma et al., 2015), banking (Sufian &
Habibullah, 2010; Wang et al., 2014). To the best of the author's knowledge,
assessing logistics efficiency at a macro level is quite limited. Markovits-
Somogyi and Bokor (2014) evaluated the logistics efficiency of European
countries using the data envelopment analysis with pair-wise comparison
(DEA-PC). In their model, they used logistics service quality and competence and timeliness (two indicators of LPI) as outputs, together with road
transport performance. Martí et al. (2017) used DEA to measure logistics
efficiency of 141 countries using the six indicators of LPI.

The rest of this chapter is structured as follows. In the following section,
the method used, i.e., the DEA, is briefly described. Section 3.3 reveals the
data used, while Section 3.4 presents the results of the study. Lastly, Section
3.5 provides a conclusion and future research directions.

3.2 METHOD

The non-parametric DEA is a tool employed to assess the efficiency of a
decision-making unit (DMU). In particular, it assesses the ability of a DMU
to obtain maximum outputs from given inputs or, equivalently, to minimize
inputs to produce given outputs (Kumbhakar & Lovell, 2000). DEA can
deal with both constant returns-to-scale (CRS) after Charnes et al. (1978);
and variable returns-to-scale (VRS) after Banker et al. (1984). The VRS
model assumes that variable returns to scale exists at the efficient frontiers,
whereas the CRS model is based on the assumption of the constant return to
scale. In addition, there are two different specifications, i.e., input-oriented
(IO) and output-oriented (OO). In the IO specification, DMUs minimize
inputs while maintaining the same level of output. On the other hand, in the
OO specification, DMUs maximize their outputs as keeping inputs constant.
In this study, the CRS-IO specification is employed following Martí et al.
(2017) since countries are assumed to strive minimizing their inputs and
cannot easily increase their outputs at least in the short term.

The production possibility set P_{CRS} is defined as

$$P_{CRS} = \left\{ (x,y) \middle| x \geq X\lambda, y \leq Y\lambda, \lambda \geq 0 \right\}, \tag{3.1}$$

where (x, y) is the observed activity belonging to P, s is the number of outputs, m is the number of inputs, N is the number of DMUs, X is a $m \times N$ matrix of inputs, Y is a $s \times N$ matrix of outputs, and λ is the non-negative intensity vector. The CRS-DEA with IO specification evaluates the efficiency of DMU_o ($o = 1, 2, \ldots, N$) by solving the following linear program:

$$
\begin{aligned}
\min \quad & \theta \\
\text{subjected to} \quad & \theta x_o - X\lambda \geq 0 \\
& Y\lambda \geq y_o \\
& \lambda \geq 0
\end{aligned}
\tag{3.2}
$$

where θ $(0 \leq \theta \leq 1)$ is the efficiency score of the observed DMU.

Let the optimal solution of Equation (3.2) is θ^*. Using this value, we solve the following linear program:

$$
\begin{aligned}
\max \quad & es^- + es^+ \\
\text{subjected to} \quad & s^- = \theta^* x_o - X\lambda \\
& s^+ = Y\lambda - y_o \\
& \lambda, s^-, s^+ \geq 0
\end{aligned}
\tag{3.3}
$$

where e is a vector of ones, $s^- \in \mathbb{R}^m$ is the input excesses and $s^+ \in \mathbb{R}^s$ is the output shortfalls. Both of them are called the slack. Let the optimal solution of Equation (3.3) be $(\lambda^*, s^{-*}, s^{+*})$. If the optimal solutions of the two linear programs above satisfy $\theta^* = 1$ and $s^{-*} = 0$, $s^{+*} = 0$, then the DMU_o is called CRS-efficient. Otherwise, the DMU_o is called CRS-inefficient. See Cooper et al. (2006) for more comprehensive discussion of the DEA model.

3.3 DATA

The data is taken from the recent 2018 LPI data. The World Bank's LPI is based on a global survey of logistics professionals which collected input on the logistics "friendliness" (i.e., how easy or difficult they experience logistics) of the nations they operate in and trade with. In the 2018 LPI edition, almost 6,000 country evaluations were conducted by logistics professionals, which covers 100 domestic LPI countries and 160 international LPI countries. This study only covers the 160 countries surveyed in the international LPI as it is intended for international comparisons.

The six indicators are: (1) customs and border management clearance's efficiency (CUSTOMS); (2) trade and transportation infrastructure's quality (INFRASTRUCTURE); (3) the simplicity of arranging competitively priced shipments (SHIPMENTS); (4) logistics services' competency and quality (SERVICE); (5) the frequency with which shipments reach consignees within

Table 3.1 Descriptive statistics

Variables	N	Mean	Median	Std. Dev.	Min.	Max.
Inputs:						
CUSTOMS	160	2.673	2.577	0.578	1.571	4.092
INFRA-STRUCTURE	160	2.723	2.547	0.674	1.556	4.374
SERVICE	160	2.815	2.699	0.610	1.883	4.311
Outputs:						
SHIPMENTS	160	2.830	2.748	0.515	1.804	3.995
TIMELINESS	160	3.236	3.172	0.576	2.037	4.410
TRACKING-TRACING	160	2.901	2.784	0.613	1.636	4.323

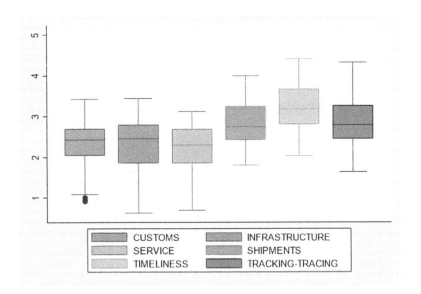

Figure 3.1 The distributions of inputs and outputs.

scheduled or expected delivery times (TIMELINESS); and (6) the ability to track and trace consignments (TRACKING-TRACING). Score for each indicator ranges from 1 to 5, the highest score represents the best logistics performance.

These indicators are mapped into two categories (Arvis et al., 2018): (1) the inputs to the supply chain (i.e., CUSTOMS, INFRASTRUCTURE, and SERVICE), and (2) supply chain performance outcomes (corresponding to LPI indicators of SHIPMENTS, TIMELINESS, and TRACKING-TRACING).

The descriptive statistics of the inputs and outputs used in this study are shown in Table 3.1. The distributions for each input and output are shown as box plots in Figure 3.1. Of the 160 surveyed countries, Germany achieved the best scores in all inputs, Belgium is the best performer in terms of SHIPMENTS and TIMELINESS, and Finland obtained the highest score in TRACKING-TRACING. On the other hand, the worst performers are: Angola for CUSTOMS, Guinea for INFRASTRUCTURE and TIMELINESS, Papua New Guinea for SERVICE, Bhutan for SHIPMENTS, and Libya for TRACKING-TRACING. The complete information can be seen in the Appendix to this chapter.

3.4 RESULTS

The input-oriented DEA with CRS environment is used to measure the efficiency of each country surveyed in the LPI 2018 edition. Notice that the inputs are scaled assuming a monotone decreasing transformation (i.e., the maximum score of five minus the actual score, see Martí et al., 2017). Table 3.2 shows the results for the twenty best and worst performers according to their logistics performance. The complete information can be seen in the Appendix. It can be observed that the best performers are mainly high-income countries. On the other hand, the twenty worst performers are mainly located in the Africa continent (e.g., Niger, Angola, Burundi) along with some other low-income countries such as Afghanistan, Bhutan, and Iraq.

According to these efficiency scores, the twenty best performers all belong to the high-income economies according to the World Bank (as of fiscal year 2023). Most of them belong to the OECD; only Singapore, Hong Kong SAR, and United Arab Emirates are non-OECD countries. On the other hand, most of the twenty worst performers belong to the low-income and lower-middle income economies, only Libya, Gabon, Cuba, and Iraq belong to the upper-middle income economies. The distributions of the efficiency score, by country, are shown in Figure 3.2. As one can observe, the high-income economies tend to have higher efficiency score than other economies. I then formally test whether the income might influence the efficiency score. The one-way analysis of variance (ANOVA) is employed to investigate whether there are significant differences that can be accrued to this particular factor. Table 3.3 shows the standard ANOVA table. The result of the ANOVA shows that the null hypothesis, i.e., the average logistics performance measure by the efficiency is equal independently of the income, may be rejected. The p-value casts doubt on the null hypothesis and suggests that at least the logistics performance in some group of countries (according to their incomes) is significantly different from other groups at the level of 5%.

The pair-wise analysis (post-hoc test) is then conducted to examine in what sense a group can be characterized by its better or worse performance. In this study, countries are assumed to come from "different populations" (i.e., they have different levels of income), thus, unlike the traditional post-hoc test of

Table 3.2 The 20 best and worst countries according to the CRS-IO DEA

Country	Efficiency score	DEA rank	LPI rank	Country	Efficiency score	DEA rank	LPI rank
20 best performers							
Germany	1.000	1	1	United Kingdom	0.729	11	9
Sweden	0.969	2	2	Australia	0.725	12	18
Japan	0.874	3	5	United States	0.715	13	14
Denmark	0.842	4	8	New Zealand	0.681	14	15
Belgium	0.821	5	3	United Arab Emirates	0.661	15	11
Netherlands	0.812	6	6	Spain	0.653	16	17
Singapore	0.802	7	7	Switzerland	0.645	17	13
Finland	0.781	8	10	France	0.608	18	16
Austria	0.768	9	4	Canada	0.586	19	20
Hong Kong SAR, China	0.748	10	12	Norway	0.571	20	21
20 worst performers							
Mauritania	0.198	141	135	Syrian Arab Republic	0.175	151	138
Senegal	0.196	142	141	Venezuela	0.174	152	142
Central African Republic	0.196	143	151	Eritrea	0.171	153	155
Chad	0.196	144	123	Haiti	0.170	154	153
Papua New Guinea	0.189	145	148	Zimbabwe	0.164	155	152
Libya	0.188	146	154	Sierra Leone	0.161	156	156
Gabon	0.181	147	150	Burundi	0.157	157	158
Cuba	0.180	148	146	Angola	0.156	158	159
Bhutan	0.179	149	149	Afghanistan	0.152	159	160
Iraq	0.178	150	147	Niger	0.149	160	157

ANOVA, the assumption of equivalence of variance does not hold. The Games-Howell test is used to do such comparisons. It is an improved version of the Tukey-Kramer method (when the assumption of equivalence of variance holds) and is applicable in cases where the equivalence of variance assumption is violated (Lee & Lee, 2018). The result is shown in Table 3.4. We can find that the differences between high-income countries with all levels of income are statistically significant at the level of 5%. Therefore, we

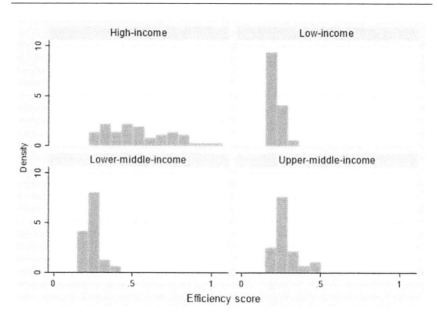

Figure 3.2 The distributions of efficiency score, by income.

Table 3.3 ANOVA: efficiency score by income

Source of variation	Sums of squares	Degrees of freedom	Mean squares	F-value	p-value
Between groups	2.868	3	0.956	60.76	0.000**
Within groups	2.439	155	0.016		
Total	5.306	158			

** significant at the level of 5%

Table 3.4 Post-hoc test: efficiency score by income

Income	Mean difference	Std. Error	p-value
Low vs Lower-middle	−0.034	0.117	0.026**
Low vs Upper-middle	−0.066	0.144	0.000**
Low vs High	−0.321	0.029	0.000**
Lower-middle vs Upper-middle	−0.032	0.136	0.098*
Lower-middle vs High	−0.287	0.029	0.000**
Upper-middle vs High	−0.255	0.030	0.000**

* significant at the level of 10%
** significant at the level of 5%

can conclude that high-income countries perform better than other levels of income in terms of logistics performance measured by the logistics performance index. A similar condition also happens in the low-income countries. However, there is a slight difference between upper-middle- and lower-middle-income countries (it is only significant at the level of 10%).

The subsequent analysis examines whether the efficiency score is influenced by the geographical area. In this sense, the World Bank's classification of region is used. The observed countries are classified into seven regions, namely, Europe and Central Asia (ECA), East Asia and Pacific (EAP), Middle East and North Africa (MENA), Latin America and the Caribbean (LAC), South Asia (SA), North America (NA), and Sub-Saharan Africa (SSA). In a similar manner, the ANOVA is used to test the null hypothesis of whether the average logistics performance measure by the efficiency is equal independently of the geographical area. The result is shown in Table 3.5. It suggests that at least the logistics performance in some group of countries (according to their regions) is significantly different from other groups at the level of 5%.

To examine which group is "different" to other groups, the post-hoc test of Games-Howell is again employed. The result is shown in Table 3.6. We can observe, for example, countries in EAP have better performance than countries in LAC (the mean difference is positive, and the difference is statistically significant at the level of 5%). However, the performance of countries in this region is not as good as that in the countries in ECA since the mean difference is not statistically significant. Among twenty-one comparisons, there are seven statistically significant pair-wise, i.e., EAP with LAC, EAP with SA, EAP with SSA, ECA with LAC, ECA with MENA, ECA with SA, and ECA with SSA. It seems that differentiating countries in terms of geographical area to predict the logistics performance is not quite meaningful.

Table 3.5 ANOVA table: efficiency score by geographical area

Source of variation	Sums of squares	Degrees of freedom	Mean squares	F-value	p-value
Between groups	1.691	6	0.282	11.834	0.000**
Within groups	3.644	153	0.024		
Total	5.335	159			

**significant at the level of 5%

Table 3.6 Post-hoc test: efficiency score by geographical area

Geographical area	Mean difference	Std. Error	p-value
EAP vs ECA	−0.018	0.057	1.000
EAP vs LAC	0.169	0.049	0.031**
EAP vs MENA	0.140	0.055	0.172
EAP vs NA	−0.217	0.080	0.365
EAP vs SA	0.197	0.053	0.014**
EAP vs SSA	0.201	0.048	0.006**
ECA vs LAC	0.187	0.034	0.000**
ECA vs MENA	0.158	0.041	0.006**
ECA vs NA	−0.199	0.072	0.424
ECA vs SA	0.215	0.039	0.000**
ECA vs SSA	0.220	0.033	0.000**
LAC vs MENA	−0.029	0.030	0.957
LAC vs NA	−0.386	0.066	0.234
LAC vs SA	0.028	0.027	0.930
LAC vs SSA	0.033	0.016	0.418
MENA vs NA	−0.357	0.070	0.213
MENA vs SA	0.057	0.036	0.679
MENA vs SSA	0.062	0.029	0.353
NA vs SA	0.415	0.069	0.190
NA vs SSA	0.419	0.065	0.222
SA vs SSA	0.005	0.025	1.000

** significant at the level of 5%

Finally, the result of the DEA (i.e., the efficiency score) is compared to the LPI score.[1] Two analyses are performed. The first uses the Pearson correlation coefficient to measure the association between the efficiency score and the LPI score. Results show that there is a strong positive correlation between these two scores ($r = 0.944$). The second analysis is performed by using the Spearman's rank correlation coefficient to measure the association between the DEA rank and the LPI rank. Results show the correlation is also strong and positive ($r_s = 0.979$).

3.5   CONCLUDING REMARKS AND FUTURE RESEARCH DIRECTIONS

By applying the DEA, the performance of countries participated in the LPI 2018 edition is estimated. It offers an interesting insight into the benchmark position of the countries regarding logistics performance. Results reveal that Germany has the (relatively) best performance among other countries while Niger is considered the worst performer. This study also examines whether the levels of income and country's geographical area are good predictors of the efficiency score. On the one side, the findings suggest that high-income countries perform better than other levels of income as the twenty best performers all belong to the high-income economies. On the other side, it is hard to conclude that countries in a particular geographical area perform better than countries in other geographical areas (see Table 3.5).

For the future research, it is encouraged to identify the determinants of (i.e., the factors that might influence) the efficiency score. One may not only want to know the levels of efficiency for each country, but also the factors that can explain that. Another interesting area for future research is to perform the analysis to the previous LPI edition (i.e., the panel data). A key advantage of using panel data is that it contains more information than a single cross-sectional data since the observations are observed repeatedly. One can take into account the heterogeneity that may exist which cannot be measured if one uses cross-sectional data. Even though DEA cannot handle panel data, its parametric counterpart, i.e., SFA, has this kind of ability. Moreover, the SFA in the panel data analysis could identify whether the efficiency has been persistent over time or time-varying (Kumbhakar et al., 2014). Distinguishing between time-varying and persistent efficiency is important since they may have different policy implications (Lai & Kumbhakar, 2018).

APPENDIX

No.	Country	Region	Income	DEA rank	Efficiency score	LPI rank	LPI score	CUSTOMS	INFRASTR-UCTURE	SERVICE	SHIP-MENTS	TRACKING-TRACING	TIME-LINESS
1	Afghanistan	SA	LI	159	0.152	160	1.949	3.265	3.193	3.081	2.104	1.697	2.382
2	Albania	ECA	UMI	99	0.250	88	2.660	2.653	2.706	2.441	2.824	2.666	3.203
3	Algeria	MENA	LMI	139	0.199	117	2.448	2.870	2.579	2.609	2.387	2.605	2.761
4	Angola	SSA	LMI	158	0.156	159	2.046	3.429	3.143	2.996	2.202	2.004	2.595
5	Argentina	LAC	UMI	83	0.270	61	2.887	2.582	2.226	2.223	2.924	3.047	3.369
6	Armenia	ECA	UMI	93	0.257	92	2.608	2.426	2.516	2.497	2.649	2.511	2.896
7	Australia	EAP	HI	12	0.725	18	3.751	1.133	1.032	1.291	3.247	3.817	3.976
8	Austria	ECA	HI	9	0.768	4	4.026	1.286	0.818	0.916	3.878	4.087	4.251
9	Bahamas, The	LAC	HI	96	0.254	112	2.525	2.316	2.592	2.732	2.501	2.518	2.751
10	Bahrain	MENA	HI	60	0.305	59	2.935	2.330	2.275	2.143	3.024	3.013	3.285
11	Bangladesh	SA	LMI	119	0.224	100	2.577	2.700	2.611	2.517	2.563	2.785	2.924
12	Belarus	ECA	UMI	101	0.248	103	2.575	2.650	2.559	2.358	2.309	2.536	3.177
13	Belgium	ECA	HI	5	0.821	3	4.039	1.337	1.016	0.869	3.995	4.051	4.410
14	Benin	SSA	LMI	67	0.290	76	2.750	2.438	2.503	2.498	2.734	2.748	3.416
15	Bhutan	SA	LMI	149	0.179	149	2.169	2.864	3.092	2.654	1.804	2.346	2.486
16	Bolivia	LAC	LMI	121	0.222	131	2.358	2.681	2.848	2.789	2.535	2.132	2.742
17	Bosnia and Herzegovina	ECA	UMI	71	0.282	72	2.809	2.368	2.579	2.198	2.842	2.895	3.211
18	Brazil	LAC	UMI	69	0.288	56	2.986	2.594	2.073	1.910	2.881	3.111	3.510
19	Brunei Darussalam	EAP	HI	77	0.276	80	2.707	2.378	2.539	2.290	2.513	2.747	3.174
20	Bulgaria	ECA	UMI	43	0.369	52	3.034	2.062	2.237	2.119	3.234	3.015	3.313
21	Burkina Faso	SSA	LI	87	0.265	91	2.621	2.587	2.569	2.540	2.918	2.402	3.038
22	Burundi	SSA	LI	157	0.157	158	2.064	3.313	3.047	2.668	2.214	2.007	2.165
23	Cambodia	EAP	LMI	100	0.250	98	2.579	2.630	2.855	2.592	2.794	2.515	3.155

	Country	Region	Income										
24	Cameroon	SSA	LMI	85	0.266	95	2.596	2.540	2.428	2.403	2.873	2.471	2.566
25	Canada	NA	HI	19	0.586	20	3.727	1.396	1.250	1.102	3.382	3.810	3.961
26	Central African Republic	SSA	LI	143	0.196	151	2.148	2.762	3.071	3.071	2.305	2.105	2.333
27	Chad	SSA	LI	144	0.196	123	2.417	2.846	2.628	2.380	2.370	2.372	2.620
28	Chile	LAC	HI	33	0.455	34	3.317	1.726	1.790	1.875	3.272	3.204	3.797
29	China	EAP	UMI	29	0.485	26	3.605	1.714	1.247	1.405	3.536	3.648	3.840
30	Colombia	LAC	UMI	55	0.315	58	2.942	2.388	2.333	2.134	3.194	3.085	3.171
31	Comoros	SSA	LMI	89	0.264	107	2.557	2.375	2.750	2.786	2.495	2.928	2.805
32	Congo, Dem. Rep.	SSA	LI	131	0.212	120	2.428	2.630	2.884	2.506	2.366	2.508	2.687
33	Congo, Rep.	SSA	LMI	102	0.247	115	2.486	2.727	2.934	2.719	2.866	2.378	2.949
34	Costa Rica	LAC	UMI	78	0.275	73	2.792	2.372	2.506	2.295	2.777	2.958	3.155
35	Côte d'Ivoire	SSA	LMI	50	0.340	50	3.082	2.221	2.113	1.773	3.207	3.136	3.227
36	Croatia	ECA	HI	45	0.367	49	3.104	2.021	1.987	1.904	2.929	3.013	3.594
37	Cuba	LAC	UMI	148	0.180	146	2.197	2.970	2.958	2.798	2.270	2.147	2.462
38	Cyprus	ECA	HI	39	0.384	45	3.151	1.948	2.108	1.995	3.148	3.148	3.623
39	Czech Republic	ECA	HI	25	0.521	22	3.680	1.713	1.535	1.284	3.746	3.703	4.134
40	Denmark	ECA	HI	4	0.842	8	3.992	1.082	1.041	0.992	3.530	4.176	4.408
41	Djibouti	MENA	LMI	104	0.245	90	2.635	2.653	2.208	2.750	2.455	2.850	3.150
42	Dominican Republic	LAC	UMI	98	0.251	87	2.662	2.595	2.643	2.558	2.770	2.975	2.981
43	Ecuador	LAC	UMI	63	0.300	62	2.882	2.198	2.278	2.249	2.751	3.071	3.190
44	Egypt, Arab Rep.	MENA	LMI	79	0.275	67	2.825	2.400	2.182	2.182	2.792	2.724	3.194
45	El Salvador	LAC	LMI	107	0.237	101	2.576	2.699	2.751	2.439	2.711	2.472	3.099
46	Equatorial Guinea	SSA	UMI	124	0.219	136	2.318	3.090	3.125	2.750	2.875	2.125	2.750

(Continued)

No.	Country	Region	Income	DEA rank	Efficiency score	LPI rank	LPI score	CUSTOMS	INFRASTR-UCTURE	SERVICE	SHIP-MENTS	TRACKING-TRACING	TIME-LINESS
47	Eritrea	SSA	LI	153	0.171	155	2.087	2.866	3.138	2.833	2.088	2.167	2.082
48	Estonia	ECA	HI	32	0.468	36	3.312	1.678	1.901	1.852	3.262	3.207	3.799
49	Fiji	EAP	UMI	136	0.203	133	2.352	2.588	2.597	2.692	2.162	2.308	2.538
50	Finland	ECA	HI	8	0.781	10	3.969	1.185	0.997	1.113	3.563	4.323	4.280
51	France	ECA	HI	18	0.608	16	3.844	1.410	1.003	1.162	3.545	3.999	4.152
52	Gabon	SSA	UMI	147	0.181	150	2.162	3.044	2.908	2.925	2.099	2.067	2.672
53	Gambia, The	SSA	LI	125	0.218	127	2.401	2.923	3.181	2.795	2.705	2.805	2.705
54	Georgia	ECA	UMI	109	0.236	119	2.443	2.576	2.618	2.743	2.376	2.257	2.945
55	Germany	ECA	HI	1	1.000	1	4.201	0.908	0.626	0.689	3.859	4.239	4.392
56	Ghana	SSA	LMI	111	0.234	106	2.565	2.548	2.557	2.493	2.534	2.571	2.871
57	Greece	ECA	HI	47	0.360	42	3.205	2.161	1.828	1.944	3.303	3.175	3.662
58	Guatemala	LAC	UMI	117	0.226	125	2.415	2.842	2.804	2.751	2.332	2.416	3.111
59	Guinea	SSA	LI	118	0.226	145	2.201	2.553	3.444	2.928	2.322	2.697	2.037
60	Guinea-Bissau	SSA	LI	138	0.199	129	2.387	2.992	3.220	2.716	2.534	2.784	2.856
61	Guyana	LAC	UMI	120	0.224	132	2.358	2.446	2.911	2.764	2.168	2.439	2.647
62	Haiti	LAC	LMI	154	0.170	153	2.112	2.968	3.055	2.814	2.005	2.054	2.439
63	Honduras	LAC	LMI	115	0.227	93	2.604	2.761	2.527	2.277	2.662	2.677	2.834
64	Hong Kong SAR, China	EAP	HI	10	0.748	12	3.920	1.185	1.031	1.068	3.768	3.918	4.139
65	Hungary	ECA	HI	31	0.478	31	3.419	1.645	1.729	1.787	3.222	3.671	3.786
66	Iceland	ECA	HI	38	0.416	40	3.225	2.231	1.814	1.395	2.792	3.353	3.697
67	India	SA	LMI	41	0.371	44	3.177	2.035	2.095	1.872	3.212	3.319	3.497
68	Indonesia	EAP	LMI	54	0.326	46	3.150	2.327	2.105	1.900	3.228	3.300	3.670
69	Iran, Islamic Rep.	MENA	LMI	65	0.292	64	2.853	2.375	2.233	2.162	2.757	2.767	3.356
70	Iraq	MENA	UMI	150	0.178	147	2.176	3.161	2.967	3.092	2.325	2.192	2.720
71	Ireland	ECA	HI	28	0.490	29	3.510	1.642	1.707	1.404	3.424	3.623	3.756

72	Israel	MENA	HI	35	0.445	37	3.308	1.683	1.671	1.611	2.783	3.500	3.591
73	Italy	ECA	HI	21	0.558	19	3.739	1.528	1.147	1.345	3.512	3.855	4.127
74	Jamaica	LAC	UMI	113	0.231	113	2.519	2.584	2.676	2.464	2.534	2.481	2.793
75	Japan	EAP	HI	3	0.874	5	4.026	1.006	0.752	0.912	3.592	4.049	4.254
76	Jordan	MENA	UMI	90	0.262	84	2.688	2.509	2.282	2.455	2.443	2.772	3.183
77	Kazakhstan	ECA	UMI	58	0.312	71	2.810	2.336	2.454	2.423	2.734	2.777	3.525
78	Kenya	SSA	LMI	74	0.280	68	2.815	2.346	2.446	2.190	2.622	3.069	3.176
79	Korea, Rep.	EAP	HI	26	0.507	25	3.612	1.597	1.274	1.412	3.330	3.754	3.920
80	Kuwait	MENA	HI	59	0.306	63	2.861	2.275	1.978	2.203	2.627	2.658	3.368
81	Kyrgyz Republic	ECA	LMI	82	0.270	108	2.546	2.250	2.625	2.642	2.215	2.644	2.941
82	Lao PDR	EAP	LMI	84	0.268	82	2.700	2.387	2.559	2.351	2.716	2.914	2.843
83	Latvia	ECA	HI	64	0.293	70	2.810	2.203	2.017	2.307	2.745	2.788	2.879
84	Lebanon	MENA	LMI	97	0.252	79	2.717	2.615	2.363	2.530	2.804	2.804	3.180
85	Lesotho	SSA	LMI	132	0.211	139	2.277	2.639	3.042	2.968	2.208	2.366	2.699
86	Liberia	SSA	LI	126	0.217	143	2.229	3.094	3.086	2.861	2.082	2.048	3.246
87	Libya	MENA	UMI	146	0.188	154	2.106	3.046	2.751	2.955	1.989	1.636	2.770
88	Lithuania	ECA	HI	48	0.350	54	3.018	2.154	2.270	2.044	2.790	3.123	3.647
89	Luxembourg	ECA	HI	22	0.548	24	3.630	1.472	1.369	1.242	3.371	3.615	3.904
90	Macedonia, FYR	ECA	UMI	91	0.262	81	2.705	2.547	2.526	2.257	2.837	2.643	3.035
91	Madagascar	SSA	LI	133	0.210	128	2.389	2.684	2.840	2.670	2.185	2.610	2.733
92	Malawi	SSA	LI	105	0.239	97	2.586	2.574	2.824	2.324	2.551	2.667	2.975
93	Malaysia	EAP	UMI	40	0.375	41	3.221	2.102	1.853	1.702	3.348	3.148	3.464
94	Maldives	SA	UMI	88	0.264	86	2.665	2.601	2.282	2.714	2.659	2.600	3.323
95	Mali	SSA	LI	112	0.231	96	2.590	2.846	2.697	2.550	2.703	3.077	2.827
96	Malta	ECA	HI	76	0.276	69	2.814	2.302	2.100	2.200	2.700	2.800	3.005
97	Mauritania	SSA	LMI	141	0.198	135	2.331	2.800	2.737	2.808	2.192	2.468	2.677

(Continued)

No.	Country	Region	Income	DEA rank	Efficiency score	LPI rank	LPI score	CUSTOMS	INFRASTR-UCTURE	SERVICE	SHIP-MENTS	TRACKING-TRACING	TIME-LINESS
98	Mauritius	SSA	UMI	75	0.280	78	2.733	2.295	2.200	2.144	2.120	2.999	2.999
99	Mexico	LAC	UMI	53	0.327	51	3.051	2.230	2.153	1.980	3.103	3.005	3.530
100	Moldova	ECA	UMI	106	0.238	116	2.456	2.746	2.980	2.700	2.693	2.206	3.167
101	Mongolia	EAP	LMI	114	0.228	130	2.373	2.776	2.904	2.792	2.485	2.100	3.061
102	Montenegro	ECA	UMI	72	0.282	77	2.746	2.439	2.425	2.276	2.684	2.575	3.329
103	Morocco	MENA	LMI	116	0.227	109	2.540	2.672	2.565	2.511	2.575	2.513	2.878
104	Myanmar	EAP	LMI	128	0.212	137	2.298	2.833	3.005	2.721	2.199	2.202	2.908
105	Nepal	SA	LMI	108	0.236	114	2.513	2.710	2.806	2.536	2.361	2.646	3.101
106	Netherlands	ECA	HI	6	0.812	6	4.019	1.082	0.792	0.912	3.682	4.025	4.253
107	New Zealand	EAP	HI	14	0.681	15	3.876	1.291	1.009	0.984	3.429	3.916	4.255
108	Niger	SSA	LI	160	0.149	157	2.070	3.235	2.999	2.900	2.001	2.223	2.334
109	Nigeria	SSA	LMI	134	0.209	110	2.532	3.033	2.440	2.601	2.522	2.685	3.068
110	Norway	ECA	HI	20	0.571	21	3.697	1.480	1.309	1.311	3.429	3.944	3.943
111	Oman	MENA	HI	44	0.369	43	3.197	2.133	1.845	1.946	3.299	2.971	3.804
112	Pakistan	SA	LMI	127	0.215	122	2.419	2.878	2.803	2.413	2.629	2.265	2.663
113	Panama	LAC	HI	46	0.365	38	3.276	2.134	1.870	1.667	3.315	3.400	3.597
114	Papua New Guinea	EAP	LMI	145	0.189	148	2.174	2.679	3.029	3.117	2.147	2.258	2.436
115	Paraguay	LAC	UMI	61	0.302	74	2.782	2.360	2.453	2.278	2.693	2.613	3.445
116	Peru	LAC	UMI	70	0.288	83	2.693	2.471	2.719	2.579	2.844	2.555	3.446
117	Philippines	EAP	LMI	57	0.313	60	2.904	2.471	2.274	2.224	3.293	3.059	2.984
118	Poland	ECA	HI	27	0.495	28	3.540	1.747	1.791	1.420	3.678	3.506	3.954
119	Portugal	ECA	HI	24	0.528	23	3.643	1.829	1.753	1.294	3.826	3.719	4.126
120	Qatar	MENA	HI	36	0.441	30	3.474	2.000	1.625	1.581	3.750	3.562	3.704
121	Romania	ECA	HI	56	0.315	48	3.119	2.419	2.093	1.926	3.176	3.265	3.682
122	Russian Federation	ECA	UMI	86	0.265	75	2.757	2.580	2.225	2.251	2.644	2.646	3.313
123	Rwanda	SSA	LI	49	0.342	57	2.975	2.333	2.245	2.149	3.391	2.751	3.351

124	São Tomé and Príncipe	SSA	LMI	80	0.273	89	2.653	2.286	2.670	2.354	2.416	2.783	3.015
125	Saudi Arabia	MENA	HI	62	0.300	55	3.011	2.339	1.893	2.140	2.985	3.172	3.300
126	Senegal	SSA	LMI	142	0.196	141	2.252	2.831	2.776	2.893	2.363	2.107	2.518
127	Serbia	ECA	UMI	66	0.291	65	2.841	2.403	2.400	2.295	2.971	2.789	3.333
128	Sierra Leone	SSA	LI	156	0.161	156	2.078	3.182	3.182	3.000	2.184	2.273	2.338
129	Singapore	EAP	HI	7	0.802	7	3.996	1.113	0.936	0.900	3.580	4.080	4.320
130	Slovak Republic	ECA	HI	51	0.330	53	3.027	2.211	2.000	1.861	3.101	2.985	3.139
131	Slovenia	ECA	HI	30	0.483	35	3.315	1.581	1.738	1.948	3.188	3.267	3.695
132	Solomon Islands	EAP	LMI	68	0.289	104	2.571	2.226	2.793	2.267	2.205	2.373	3.117
133	Somalia	SSA	LI	135	0.205	144	2.209	3.000	3.187	2.695	2.613	2.233	2.204
134	South Africa	SSA	UMI	34	0.452	33	3.376	1.825	1.813	1.807	3.508	3.411	3.742
135	Spain	ECA	HI	16	0.653	17	3.831	1.379	1.160	1.200	3.830	3.835	4.063
136	Sri Lanka	SA	LMI	103	0.247	94	2.598	2.418	2.511	2.576	2.515	2.787	2.787
137	Sudan	SSA	LMI	130	0.212	121	2.428	2.864	2.818	2.495	2.582	2.505	2.616
138	Sweden	ECA	HI	2	0.969	2	4.053	0.951	0.760	1.023	3.916	3.876	4.285
139	Switzerland	ECA	HI	17	0.645	13	3.901	1.371	0.978	1.033	3.515	4.096	4.242
140	Syrian Arab Republic	MENA	LI	151	0.175	138	2.296	3.181	2.486	2.711	2.366	2.366	2.443
141	Taiwan, China	EAP	HI	23	0.536	27	3.600	1.526	1.280	1.435	3.477	3.671	3.721
142	Tajikistan	ECA	LMI	140	0.198	134	2.340	3.077	2.834	2.667	2.313	2.333	2.949
143	Thailand	EAP	UMI	37	0.438	32	3.411	1.858	1.862	1.589	3.457	3.467	3.814
144	Togo	SSA	LI	122	0.222	118	2.447	2.686	2.768	2.750	2.519	2.450	2.879
145	Trinidad and Tobago	LAC	HI	110	0.236	124	2.416	2.577	2.617	2.729	2.586	2.267	2.532
146	Tunisia	MENA	LMI	94	0.255	105	2.570	2.625	2.902	2.702	2.498	2.865	3.236

(Continued)

No.	Country	Region	Income	DEA rank	Efficiency score	LPI rank	LPI score	CUSTOMS	INFRASTR-UCTURE	SERVICE	SHIP-MENTS	TRACKING-TRACING	TIME-LINESS
147	Turkey	ECA	UMI	52	0.328	47	3.146	2.287	1.790	1.953	3.061	3.233	3.628
148	Turkmenistan	ECA	UMI	129	0.212	126	2.410	2.650	2.771	2.692	2.288	2.558	2.719
149	Uganda	SSA	LI	81	0.272	102	2.575	2.389	2.806	2.504	2.761	2.414	2.899
150	Ukraine	ECA	LMI	73	0.282	66	2.830	2.509	2.779	2.157	2.829	3.109	3.422
151	United Arab Emirates	MENA	HI	15	0.661	11	3.956	1.369	0.979	1.081	3.847	3.960	4.376
152	United Kingdom	ECA	HI	11	0.729	9	3.987	1.228	0.967	0.950	3.672	4.108	4.330
153	United States	NA	HI	13	0.715	14	3.885	1.225	0.955	1.126	3.506	4.092	4.084
154	Uruguay	LAC	HI	92	0.259	85	2.685	2.485	2.567	2.291	2.735	2.780	2.906
155	Uzbekistan	ECA	LMI	123	0.220	99	2.577	2.897	2.430	2.412	2.423	2.709	3.090
156	Venezuela, RB	LAC	*	152	0.174	142	2.229	3.213	2.903	2.793	2.380	2.294	2.582
157	Vietnam	EAP	LMI	42	0.370	39	3.274	2.050	1.995	1.601	3.155	3.450	3.672
158	Yemen, Rep.	MENA	LI	137	0.200	140	2.265	2.599	2.883	2.740	2.207	2.162	2.426
159	Zambia	SSA	LI	95	0.255	111	2.526	2.820	2.697	2.519	3.053	1.981	3.053
160	Zimbabwe	SSA	LMI	155	0.164	152	2.120	3.000	3.167	2.840	2.056	2.262	2.386

*Venezuela has been temporarily unclassified as of July 2021 pending release of revised national accounts statistics.

NOTE

1 The detailed information of how to construct the LPI 2018 score can be found in Arvis et al. (2018).

REFERENCES

Agasisti, T., & Zoido, P. (2018). Comparing the efficiency of schools through international benchmarking: Results from an empirical analysis of OECD PISA 2012 data. *Educational Researcher, 47*(6), 352–362.

Agasisti, T., & Zoido, P. (2019). The efficiency of schools in developing countries, analysed through PISA 2012 data. *Socio-Economic Planning Sciences, 68*, 100711.

Arvis, J.-F., Ojala, L., Wiederer, C., Shepherd, B., Raj, A., Dairabayeva, K., & Kiiski, T. (2018). *Connecting to Compete 2018 Trade Logistics in the Global Economy: The Logistics Performance Index and Its Indicators.* Washington, DC: World Bank Group.

Banker, R. D., Charnes, A., & Cooper, W. W. (1984). Models for the estimation of technical and scale inefficiencies in data envelopment analysis. *Management Science, 30*, 1078–1092.

Charnes, A., Cooper, W. W., & Rhodes, E. (1978). Measuring the efficiency of decision making units. *European Journal of Operational research, 2*(6), 429–444.

Cooper, W. W., Seiford, L. M., & Tone, K. (2006). *Introduction to Data Envelopment Analysis and Its Uses: With DEA-Solver Software and References.* New York, NY: Springer.

De Witte, K., & López-Torres, L. (2017). Efficiency in education: A review of literature and a way forward. *Journal of the Operational Research Society, 68*(4), 339–363.

Handayani, N. U., Sari, D. P., Ulkhaq, M. M., Widharto, Y., & Fitriani, R. C. A. (2020). A data envelopment analysis approach for assessing the efficiency of subsectors of creative industry: A case study of batik enterprises from Semarang, Indonesia. *AIP Conference Proceedings, 2217*(1), 030038.

Kumbhakar, S. C., Lien, G., & Hardaker, J. B. (2014). Technical efficiency in competing panel data models: A study of Norwegian grain farming. *Journal of Productivity Analysis, 41*(2), 321–337.

Kumbhakar, S. C., & Lovell, C. K. (2000). *Stochastic Frontier Analysis.* Cambridge: Cambridge University Press.

Lai, H.-P., & Kumbhakar, S. C. (2018). Panel data stochastic frontier model with determinants of persistent and transient inefficiency. *European Journal of Operational Research, 271*, 746–755.

Lee, S., & Lee, D. K. (2018). What is the proper way to apply the multiple comparison test?. *Korean Journal of Anesthesiology, 71*(5), 353–360.

Markovits-Somogyi, R., & Bokor, Z. (2014). Assessing the logistics efficiency of European countries by using the DEA-PC methodology. *Transport, 29*(2), 137–145.

Martí, L., Martín, J. C., & Puertas, R. (2017). A DEA-logistics performance index. *Journal of Applied Economics, 20*(1), 169–192.

Moradi, M., Nematollahi, M. A., Mousavi Khaneghah, A., Pishgar-Komleh, S. H., & Rajabi, M. R. (2018). Comparison of energy consumption of wheat production in conservation and conventional agriculture using DEA. *Environmental Science and Pollution Research*, *25*(35), 35200–35209.

Ojha, V. K., Goyal, S., & Chand, M. (2022). A review of data-driven decision making in advance manufacturing systems. *Computer Integrated Manufacturing Systems*, *28*(11), 172–182.

Ojha, V. K., Goyal, S., & Chand, M. (2023). Study on data-driven decision-making in entrepreneurship. In *Principles of Entrepreneurship in the Industry 4.0 Era* (pp. 75–88). Boca Raton: CRC Press.

Pramono, S. N. W., Ulkhaq, M. M., Pujotomo, D., & Ardhini, M. A. (2019). Assessing the efficiency of small and medium industry: An application of data envelopment analysis. *IOP Conference Series: Materials Science and Engineering*, *598*(1), 012043.

Santín, D., & Sicilia, G. (2018). Using DEA for measuring teachers' performance and the impact on students' outcomes: Evidence for Spain. *Journal of Productivity Analysis*, *49*(1), 1–15.

Sari, D. P., Handayani, N. U., Ulkhaq, M. M., Budiawan, W., Maharani, D. L., & Ardi, F. (2018). A data envelopment analysis approach for assessing the efficiency of small and medium-sized wood-furniture enterprises: A case study. *MATEC Web of Conferences*, *204*, 01015.

Sufian, F., & Habibullah, M. S. (2010). Developments in the efficiency of the Thailand banking sector: A DEA approach. *International Journal of Development Issues*, *9*(3), 226–245.

Toma, E., Dobre, C., Dona, I., & Cofas, E. (2015). DEA applicability in assessment of agriculture efficiency on areas with similar geographically patterns. *Agriculture and Agricultural Science Procedia*, *6*, 704–711.

Ulkhaq, M. M. (2022). Assessing technical efficiency of large and medium manufacturing industry in West Java Province, Indonesia: A data envelopment analysis approach. *The ES Management and Business*, *1*(1), 24–30.

Ulkhaq, M. M., & Pratiwi, T. N. (2022). A data envelopment analysis approach to assess technical efficiency of large and medium manufacturing industry in Central Java Province, Indonesia. *International Economic and Finance Review*, *1*(2), 54–65.

Wang, K., Huang, W., Wu, J., & Liu, Y. N. (2014). Efficiency measures of the Chinese commercial banking system using an additive two-stage DEA. *Omega*, *44*, 5–20.

Chapter 4

Artificial intelligence to complement Lean approach in the healthcare industry

Sayantan Chakraborty
Amity University, Gurgaon, India

Palka Mittal
Delhi Pharmaceutical Sciences & Research University, New Delhi, India

Koustuv Dalal
Mid Sweden University, Sundsvall, Sweden

Jaseela Majeed
Delhi Pharmaceutical Sciences & Research University, New Delhi, India

Luxita Sharma
Amity University, Gurgaon, India

4.1 INTRODUCTION

Globally, the healthcare sector is among the largest and most rapidly grow-ing sectors. The healthcare sector, sometimes referred to as the medical sec-tor or the health economy, is a grouping and integration of economic sectors that provide goods and services for the treatment of patients in the categoties of curative, preventive, rehabilitative, and palliative care. Healthcare services are constantly struggling to raise the standard of care, boost pro-ductivity, and ultimately give patients value for their money.

The provision of medical assistance is considered a service in the health-care industry. To increase production and efficiency, the healthcare system must be thoroughly upgraded. Because healthcare affects everyone's life, it is more than just a business. In terms of healthcare, there is minimal space for error. A few simple errors can have a catastrophic effect on hundreds of people. According to the *Journal of Healthcare Finance*, medical errors cost the US about $19.5 billion in 2008. The Institute of Medicine estimated that in 2008 a fatality occurs roughly every two minutes, 548 every day or 200,000 in a year in the US alone [1, 2].

DOI: 10.1201/9781003462163-4

There is a tacit understanding in this regard that internal inefficiencies can cause delays in care and overcrowding, which can have a negative impact on patient safety, patient/healthcare staff satisfaction, and the general standard of care. Examples of internal inefficiencies include poor patient flow and insufficient utilization of resources [3].

What can the healthcare industry do to maintain efficiency, high standards of patient safety, and employee involvement while also maintaining profitability given the daily rise in healthcare costs and the decline in reimbursement rates? The answer is the execution of amalgamated Lean principles with artificial intelligence. Lean and Artificial Intelligence (AI) can work together to increase service quality while lowering costs and waste, which has a positive impact on healthcare.

A healthcare professional who is knowledgeable about the application of tools and methods to logically solve problems and promote the standard of care is well positioned to excel in their organizations. Along with reducing operating expenses and improving patient care quality, they also save deaths from happening [4]. Lean in the healthcare industry helps in improving the lines of communication across all levels of management. As a result, all employees collaborate as a team to reduce unnecessary waste inside the organization and raise the standard of care. The implementation of Lean concepts in the healthcare industry guarantees a high level of efficiency and prompt patient care. The patient wait time is cut down, and worker safety is improved. Lean concepts also enhance hospital admittance procedures. It ensures improved efficiency with patient appointments and records, enhances diagnostics, and can make a significant difference in lowering unfortunate avoidable deaths.

The applications of AI range from improving pathology classification, such as in categorizing images in radiology or cardiology, differentiating ECG characteristics, and identifying illness patterns and epidemiology. AI can help reduce the burdensome workload that healthcare personnel is under and hence improve their efficiency [5]. The workload of healthcare professionals might be divided up using AI, which would streamline human workers' efforts. To achieve this goal, AI might be applied to prevent unnecessary hospital admissions from taking place at all. Some tasks that are typically completed by healthcare personnel can be replaced by AI. Many of the administrative tasks that doctors or nurses perform are routine and do not need much thought. AI applications could quickly take their place [6]. For instance, Ting et al. showed that human graders and AI achieved similar diagnostic results in a national program to screen for diabetic retinopathy, a finding which has the potential to reduce the screening workload by 75% [7]. Additionally, this might lower the cost of extensive screening services [8]. AI might also help in decision-making. Medical error is a major problem in healthcare today since all people, including doctors, are prone to making mistakes. It makes sense that they would make mistakes given their colossal workload and stress. Hence, it could be ideal for

AI to serve as another set of eyes, to decrease medical errors. By offering current advice on clinical recommendations or advancements, AI can facilitate better clinical decision-making in addition to compensating for unintended errors [5].

Since both Lean and AI are promising tools to improve efficiency and effectiveness by reducing errors and waste in the healthcare industry, their integration is no less than a boon.

4.2 THE LEAN CONCEPT IN HEALTHCARE

All industrial and service industries strive tirelessly to reduce costs and improve market position in a highly competitive environment. Maintaining with a competitive edge and the rising expense of healthcare services present even greater challenges for healthcare businesses. Owners and managers want to provide superior-quality medical services at economic prices [9]. To maintain quality at economic rates, many firms, including the healthcare industry, have adopted Lean as one of their quality improvement methodologies to gain an advantage.

The term "Lean" was first presented in the book, *The Machine that Changed the World,* by Womack et al. Toyota Motors first adopted Lean techniques in 1990 in manufacturing and dubbed their approach "The Toyota Way" [10].

Lean is a reengineering philosophy that seeks to increase process and system effectiveness by minimizing waste and non-value-added operations [11, 12]. Lean has drawn a lot of attention in the healthcare sector to create procedures that are risk-free, effective, and waste-free [11].

The UK and the USA made the first attempts to apply Lean thinking to healthcare in 2001 and 2002, respectively [13]. Lean execution in healthcare has grown significantly over the past 20 years, especially in poor nations [14].

Lean healthcare is built on a five-stage process [15] that is described as follows:

(i) To meet the demands, it is essential to define the value of the clients in terms of diagnostic tests, therapies, and treatment procedures.
(ii) Map the value, which entails specifying processes from the initiation to the completion of the process phases.
(iii) Analyzing the value stream to find waste and eliminate it, or, in the context of healthcare, adapting and becoming effective.
(iv) Pulling, which is the capacity to indicate the speed of the actions for the subsequent phases to prevent stockpiles.
(v) The pursuit of perfection is a stage that must drive Lean healthcare's continual improvement with timely and high-quality care [16, 17].

4.2.1 Wastes in healthcare

Transportation, inventory, motion, waiting, overprocessing, overproduction, and defect are the seven types of waste acknowledged by Taichi Ohno in terms of production. Patient care issues and dissatisfaction are referred to as waste in the healthcare industry [14]. Waste leads to disruptions in the healthcare service, inconsistent care, and delivery, which raises costs, causes errors and demotivates staff [18]. Various wastes in the healthcare industry are shown in Figure 4.1.

i. **Transportation**: Transportation in a hospital setting refers to a variety of patient movements that take place from the time a patient enters the facility until they are discharged (such as moving from one department to another, moving for a lab test, moving for medicine, etc.). Numerous hospitals have specific locations for particular equipment types, which causes unnecessary patient travel. Waste associated with transportation in healthcare includes unneeded transfers of patients, diagnostic samples, medications, and goods [14].

ii. **Inventory**: Inventory waste refers to any supply that is either unavailable or more than what is needed. To carry out the desired task in a hospital, several supplies, consumables and non-consumables, medical equipment, and medications are required. When these resources are maintained in excess, it creates additional inventory and having a lot of stock might result in cash shortages and the expiration of supplies

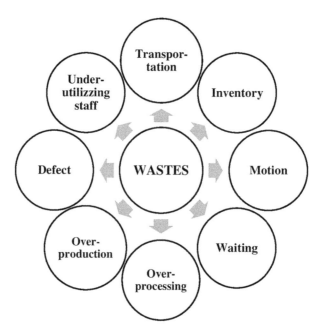

Figure 4.1 Wastes in healthcare.

and medications. An additional form of waste in the healthcare process is when patients are kept for longer than necessary, leaving the facility unavailable for another patient. Inventory management is essential for overseeing quality and assessing the efficiency of the process [19].

iii. **Motion:** Waste of motion is the term used to describe unnecessary movements made by employees or workers, in the case of the healthcare system. To reduce motion waste, the healthcare procedure must be simplified. The most prevalent motion waste in hospitals is the movement of medical staff to treat patients. In a hospital, technicians, nurses, and doctors cannot remain still, but one must concentrate on minimizing needless movements [14].

iv. **Waiting:** Waiting is the time that a patient spends inactive. Due to a delay in treatment and following processes, patients have to wait for long periods of time. Employees occasionally wait because of an uneven workload. Patients generally wait for a diagnosis from a doctor, admission to an in-patient ward, or even while discharged from the hospital. Waiting for medical procedures results in the underuse of the available resources, responsible for the inefficient procedures at the healthcare facility [20, 21].

v. **Overprocessing:** The term "overprocessing" refers to the improper use of processing. Overprocessing typically happens when a procedure is performed at a level higher than the patient has requested or required. One illustration is the requirement for duplicate pathology tests [22].

vi. **Overproduction:** It can be challenging to identify overproduction in the healthcare industry, which might happen when patients seek unnecessary operations or buy extra medications out of undue caution. Unnecessary radiological or pathological tests and superfluous follow-ups are known as overproduction waste. Sometimes inappropriate treatments were recommended, which caused patients and staff to exert excessive effort [13].

vii. **Defect:** Any activity that was not completed to the appropriate standard was referred to as a defect. Defects or mistakes are particularly serious in the healthcare industry since they might cause patients injuries or even death. Healthcare errors typically result from improper protocol, poor communication, or incorrect diagnoses [14].

viii. **Underutilizing staff:** Underuse includes characteristics such as reluctance to share knowledge or failing to take benefit of someone's experience and inventiveness [23].

4.2.2 Variables of Lean implementation in healthcare

Lean operating concept work by minimizing waste and waiting times, and providing patients with the greatest value possible. It seeks to alter organizational values and thought, which, over time, affects how the organization behaves and cultivates its culture. Variables of Lean implementation in healthcare [10, 23, 24] are shown in Table 4.1.

Table 4.1 Lean variables in healthcare

Sr. No.	Variable	Description
i.	Goal specificity	Corresponds to the administration, medical, paramedical, and all other healthcare professionals having clear goals about what is required of them and what they may anticipate from the organization. Lean principles emphasize the approach that focuses on the customer i.e., patients.
ii.	Lean leadership	The organizational leadership necessary to adopt and maintain the Lean system is referred to as Lean leadership. In a Lean-influenced atmosphere, effective leadership entails clearly articulating the organization's goal and aligning employees with it, developing a teaching and learning strategy, and involving subordinates in decision-making.
iii.	Clarity of organizational vision	To achieve organizational goals, Lean practices call on employees of organizations to create their vision.
iv.	Financial capability	An organization that wants to implement Lean must make significant investments in resource development and expansion, new software and training materials, professional guidance, an evaluation and incentive system, infrastructure improvements, etc.
v.	Professional organizational culture	Employees' attitudes, behaviors, and ways of thinking must change to create a Lean workplace. The primary necessity for implementing Lean is to establish a culture that is encouraging and supportive.
vi.	Lean training	To continuously update and improve the work process, training offers a complete awareness of available resources in a skilled way.
vii.	Competency and expertise	An expert workforce ensures the progress, advancement, and accomplishment of the organization as well as value-added patient services.
viii.	Value addition	Lean figures out the procedures that succeed and satisfies the requirements of the patients and remove non-value-added tasks.
ix.	Patient's involvement in a quality program	Patients should be included in organizational quality programs such as audits, product purchasing, etc. to make them feel appreciated. Additionally, when managers and patients collaborate to design quality improvement initiatives, the outcomes are more beneficial since real-world service user experience is taken into account.
x.	Employee involvement	Participation of staff members in decision-making and improvement processes, increase their morale and confidence, leading to greater enjoyment, motivation, and job engagement.

(Continued)

Table 4.1 (Continued)

Sr. No.	Variable	Description
xi.	Teamwork and interdepartmental cooperation	Lean culture requires effective departmental collaboration, and staff members must be aware that they must adhere to Lean principles when working.
xii.	Time constraints for Lean implementation	The deployment of Lean is seen as a time-consuming procedure because numerous other interconnected processes must be managed at the same time.
xiii.	Employee resistance to Lean culture	Refers to the belief that Lean is seen as a cost-cutting and layoff plan, who believe that participating in the Lean execution will ultimately result in their job loss.
xiv.	Communication of goals and results	Employee motivation is increased when goals and results are communicated, as excellent communication on both a vertical and horizontal level is crucial to the implementation of Lean.
xv.	Follow-up and evaluation	Employees lack clarity because they are unsure if the objectives have been met and what the outcomes of the change are due to poor follow-up and evaluation. As a result, evaluation and follow-up are crucial and must be carefully considered.

4.2.3 Factors affecting successful Lean implementation

Lean saves money when it results in shorter cycle times, less raw material consumption, fewer quality problems, and lesser personnel required to complete the task. When Lean results in space savings or increases the capacity to handle more patients, it helps healthcare businesses to generate additional revenue. Likewise, in other industries, few variables are there that impact the successful implementation of Lean variables in the healthcare industry as shown in Figure 4.2. These independent factors lead to the successful implementation of Lean and result in both patient satisfaction and financial gains [25].

  i. **Leadership:** An ideal leadership practice pertains to supportive supervision, monitoring of execution, and follow-up of all the activities during and after implementation. Effective leadership includes mentoring, teaching, and approving required training [26].
 ii. **Employee involvement:** In the healthcare industry, employee involvement talks about the involvement of physicians, nurses, paramedical staff, and administrative staff. Administrative staff are involved in policy-making but in healthcare initiatives this is not possible without the agreement of medical and paramedical staff as they are directly in contact with the consumers [27].

Figure 4.2 Variables for successful Lean implementation.

iii. **Customer focus:** Customer focus deals with the extent of interest in the customer's requirement and satisfaction shown by the organization. Healthcare should be patient-centered;, therefore, while making any change in the system patient requirements must be the clear focus. The majority of adjustments are made to reduce operating expenses or increase revenue [28].

iv. **External forces:** External influences are the commercial pressures with which healthcare organizations must contend. The healthcare company may be compelled by local rules and competitive pressure to look for cost-cutting measures in every area. However, this might make implementation less successful [25].

v. **Employee training:** Perform training for everyone, including doctors, and a quick training need analysis (TNA) may be established after the gaps have been identified. Additionally, while providing general training to everyone as awareness workshops, the Lean team may receive customized training [25].

vi. **Holistic approach:** A comprehensive strategy means implementing Lean entirely, not only with one straightforward tool or in one specific area. When implementing an initiative holistically, all organizational and individual-related elements and problems must be taken into account [29].

4.2.4 Benefits of Lean in healthcare

According to evidence-based research, the Lean management approach successfully reduces patient length of stay, emergency department wait times, and costs, improves safety and quality of care, and improves the physical working conditions for healthcare staff. Lean implementation in healthcare is said to have both tangible and intangible benefits. The tangible benefits include decreased waiting times, reduced costs, and minimum procedural errors, and the intangible benefits are improved patient motivation, enhanced healthcare quality, and higher levels of patient happiness [14].

4.3 ARTIFICIAL INTELLIGENCE AND LEAN INTEGRATION IN HEALTHCARE

Replicating human intellectual processes in machines, particularly computers, is known as artificial intelligence (AI). Speech recognition, machine learning, expert systems, and natural language processing are just a few of the applications of AI.

AI and its associated technologies are becoming increasingly common in commerce and society, and they are now being used in healthcare. Numerous facets of patient care as well as operational procedures within providers, payers, and pharmaceutical companies could be transformed by these technologies. A wide range of studies have shown that AI is capable of performing critical healthcare tasks, such as disease diagnosis, as efficiently or higher than humans. Today, computers are already more accurate than radiologists in identifying cancerous tumors and directing researchers in the development of cohorts for expensive clinical trials.

Artificial intelligence types that are pertinent to the healthcare industry include machine learning, deep learning and neural networks, natural language processing, rule-based expert systems, physical robots, and robotic process automation.

4.3.1 Diagnosis and treatment applications

Clinical workflows and HER (Electronic Health Record) systems might be difficult to integrate AI-based diagnosis and therapy suggestions into. Perhaps more of a hurdle to the broad adoption of numerous AI-based abilities for diagnosis and therapy has been limited by AI's incapacity to provide accurate and helpful recommendations [30].

4.3.2 Applications for patient involvement and adherence

Long seen as the final hurdle between ineffective and beneficial health outcomes, engagement and adherence of the patient have been called the "last mile" challenge in healthcare. Utilization, financial results, and member experience are all improved when patients take an active role in their health and treatment. Big data and AI are increasingly being used to address these issues [31].

4.3.3 Administrative applications

In addition, some healthcare companies have tried using chatbots for tele-health, mental health, and well-being. Simple transactions, such as scheduling an appointment or renewing a prescription, may benefit from the use of these natural language processing (NLP)-based applications. Machine learning, which applies to claims and payment administration, is another branch of AI that may be used to probabilistically match data across several databases [32].

4.3.4 Implications for the healthcare workforce

The lack of job impact can be partially attributed to the industry's slow adoption of AI and the challenge of incorporating it into clinical workflows and HER systems [33].

4.3.5 Ethical implications

Healthcare decisions have largely been decided by humans in the past, and using smart machines to make or assist with them raises issues of accountability, transparency, permission, and privacy [34].

4.3.6 AI in Healthcare

AI in healthcare frequently uses NLP technologies that can understand and categorize clinical documents. NLP algorithms can analyze unstructured clinical notes for patients, giving exceptional insight into quality comprehension, better methods, and improved patient results [35].

4.3.7 The future of AI in healthcare

This talent was largely responsible for the development of precision medicine, which is widely regarded to be a critically important improvement in healthcare. We also predict that AI will eventually become proficient in generating suggestions for diagnosis and therapy, despite the challenges of early attempts. The biggest challenge for AI in the healthcare industry is not whether the technology will work but rather how to ensure its acceptance in everyday clinical practice [31].

Lean is frequently viewed as a collection of methods and tools for enhancing operations. The terms "process," "value stream," and "continuous flow" are utilized in practically every piece that examines how Lean is being applied to healthcare. The main goal of Lean healthcare is to streamline processes by identifying activities that bring value and eliminating those that don't. It is frequently stressed that the fragmented nature of the current healthcare systems calls for a change in how the delivery of patient care is viewed and structured.

Lean manufacturing in healthcare largely concentrates on three major areas: establishing value from the standpoint of the patient; mapping value streams; and minimizing waste. This creates a continuous flow.

Value stream mapping is the Lean tool that is most frequently utilized in the healthcare sector. The normal implementation procedures include doing Lean training, beginning pilot projects, and putting improvements into action with interdisciplinary teams [36].

One of the obstacles is the dearth of educators and consultants with experience in the healthcare industry who can offer guidance by exchanging experiences and providing examples of real-world uses of Lean in the industry. Lean in healthcare appears to have similar facilitators to those of other change initiatives. The performance of the healthcare system and the growth of employees and the working environment can be split into two main categories of results.

Large healthcare organizations' care procedures rarely emerge from deliberate design or deliberate action; instead, they just evolve. Since the healthcare systems are designed with a focus on the doctors, nurses, and other clinical personnel, they are frequently not the best system for patients. The care is structured within departmental silos, and, frequently, only the patient can witness the entire patient journey. A patient can frequently spend hours in hospitals under such arrangements, with just about 10 minutes of that time being valuable [37]. Breaking down the silo mentality and enabling improvements to occur across functional boundaries may be possible through the application of Lean thinking, particularly value stream and continuous flow. A large portion of the work done in the healthcare environment is not of direct benefit to patients [38]. Rarely is the optimal way that procedures should operate in healthcare operations explicitly stated. The consequences of this failure include inconsistent service, erratic access to resources and procedures, and frequent disruptions, leading to inefficiency, protracted wait times, an elevated risk of mistakes, and dissatisfied employees [39].

It is essential that information on how Lean might be used in healthcare settings be shared and that businesses learn from one another's failures and triumphs. Lean healthcare is a hotly disputed topic, but the publications that are now accessible in the field only provide a partial picture of the possible benefits and drawbacks of Lean in the healthcare industry. We could not find any publications that were critical of Lean's use in healthcare. Numerous works on the subject lack empirical support and are speculative in nature [19]. To assess the true impact and learn more about the underlying elements affecting Lean's success and sustainability in the healthcare industry, a more thorough study is needed.

4.4 OPPORTUNITIES AND CHALLENGES OF ARTIFICIAL INTELLIGENCE AND LEAN INTEGRATION IN HEALTHCARE

Four concepts—relieving, splitting up, replacing, and augmenting (Figure 4.3)—can be used to analyse how AI will affect the whole workforce as well as the healthcare sector in particular [40].

Like every other significant industry, healthcare adheres to the creative imperative to continuously advance. Many people in the healthcare sector are now choosing to reinvent how they provide healthcare through Lean management, primarily due to issues with providing quality and value to the patients [42].

The Lean manufacturing philosophy was created by Toyota and has now spread to other industries, particularly the hospital and healthcare systems. Healthcare institutions have embraced the Lean methodology more frequently in recent years as a way to boost productivity, cut expenses, get rid of waste, and plan physical layouts [43].

Lean thinking and Lean production can coexist in the healthcare industry because we are more concerned with services than we are with manufacturing processes. Patient care is the only focus in workplaces that have adopted

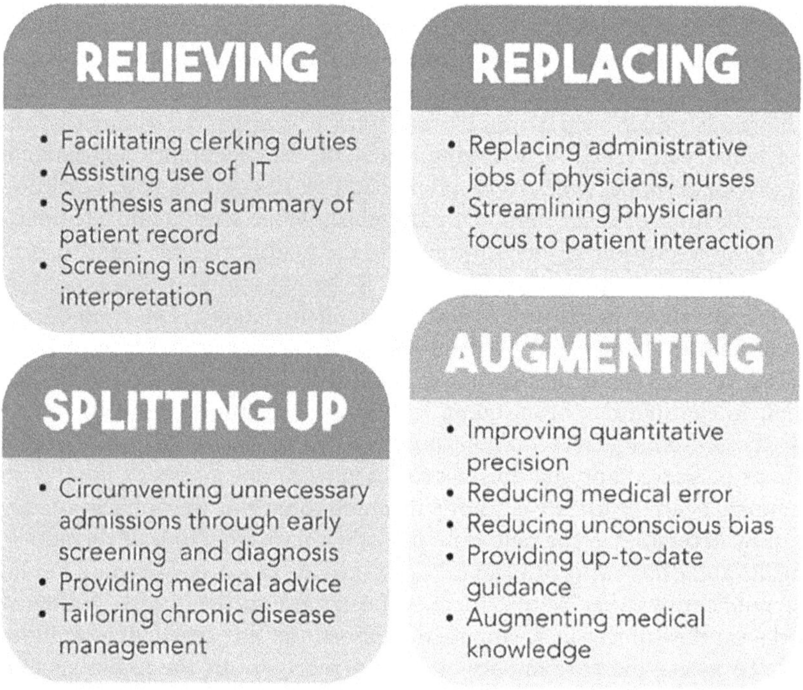

RELIEVING

- Facilitating clerking duties
- Assisting use of IT
- Synthesis and summary of patient record
- Screening in scan interpretation

REPLACING

- Replacing administrative jobs of physicians, nurses
- Streamlining physician focus to patient interaction

SPLITTING UP

- Circumventing unnecessary admissions through early screening and diagnosis
- Providing medical advice
- Tailoring chronic disease management

AUGMENTING

- Improving quantitative precision
- Reducing medical error
- Reducing unconscious bias
- Providing up-to-date guidance
- Augmenting medical knowledge

Figure 4.3 Summary of the benefits of AI in healthcare [41].

Lean management principles. This indicates that improving client happiness while doing it profitably is the top objective [19].

However, because the patient is actively involved throughout the entire process rather than merely at the receiving end like a consumer, implementing Lean in the service sector differs significantly from doing so in the manufacturing sector and is linked with various requirements. As a result, unlike in the production field, Lean components are now more subjectively measurable [44].

Other parts are occasionally more carefully examined in the service area, such as:

- Presence, accessibility, availability, tangibility
- Dependability
- Attention, helpfulness
- Reassurance (Does everything work as it should?)
- Empathy

The disillusioned notion that streamlining operations and cutting costs can result in a reduction in the quality of care provided to patients is one of the main causes of the widespread reluctance to implement Lean in the healthcare industry. The exact opposite, though, is true. Even though numerous redesign initiatives have been attempted in the past with variable and elusive degrees of success, numerous publications declare redesign and Lean methodologies to be "very productive."

According to evidence-based research, the Lean management approach successfully reduces patient length of stay, emergency department wait times, and costs, improves safety and quality of care, and improves the physical working conditions for healthcare staff [45].

Despite Lean's enormous success in the manufacturing workplace, Lean principles still need to be implemented in daily practice, which will take a lot of work.

At present, it is not so necessary that all private organizations and government organizations in healthcare have sufficient funding, for the implementation of Lean six sigma and artificial intelligence needs financial resources (i.e., Resources development and expansion, the purchase of new software, expert consultation, an evaluation and reward system, infrastructure development, etc.) [26, 46]. The success of the healthcare organization and the real worth of the patient services are ensured by the qualified and experienced staff [46]. Since an organization's position, it takes power to adequately accept and maintain the Lean system, the introduction of a Lean environment culture necessitates a change in both the employers' and employees' entire thought processes. To achieve this sustainable change, organizations must foster an environment that is supportive of their employees [45, 47]. Fulfill the patient's needs and correspond to the identification of processes that have been performed. Non-value-added tasks need to be found and removed, according to Lean thinking [47]. When managers and patients work together to plan quality improvement initiatives, the outputs are more beneficial because the experiences of real service users are incorporated, but patient involvement is not always easy. The involvement of patients in quality audits of companies also generated higher success [22]. Employee morale and productivity will grow in the Lean system if they are involved in judgment and career advancement. This will increase their happiness, motivation, and engagement in the process, which will improve their job [48]. In Lean culture there is intradepartmental coordination is required but the lack of this factor may produce difficulties in the implementation of the Lean system [49]. Since numerous other linked processes must be managed continuously, implementing Lean in between organizational tasks is challenging and is believed to be a lengthy procedure [29]. If the goal is not very clear to the employees, this can be a major disaster for the implementation of Lean six sigma integration into AI. Management, doctors, surgeons, nurses, and other health workers should all understand what is expected of them and also what they can expect from their organization. In AI and Lean system setup, the manpower that is needed is well trained by the professional trainers, but sometimes it is not possible to just provide them with proper training/knowledge. The follow-up and evaluation need to be very perfect for the sustainability of the Lean system and the lack of these two may lose clarity among the employees as they are unaware of the objectives of the organization being reached [29, 50].

The integration of the Lean approach can boost the production rate or quality production rate because the AI-incorporated integrated healthcare manufacturing can result in a very huge boost to healthcare products and the incorporated Lean approach can result in the betterment of the quality of that production. The expenses are also minimized because the Lean approach mainly works upon reducing the errors or waste so a better Lean plan can result in the reduction of the input cost; the cost of the manpower is also reduced due to the AI integration. "Human Errors" is a term that makes a huge difference when it comes to the healthcare industry because there is zero tolerance for these kinds of errors when you hold a life in your hand; they can be much reduced as a result of the incorporation of AI and Lean techniques in healthcare. The efficacy and efficiency of the processes and Standard Operating Procedures (SOPs) are also much improved due to this incorporation. Every industry needs to make maximum output from the minimum input; these types of goals can be achieved by this amalgamated Lean approach (the integration of Lean and AI). The healthcare research sectors also benefits from this amalgamated Lean approach. The management of the resources is also simplified and the loopholes are easily identified with the help of this approach. The process of monitoring and evaluation can be modified and improved with the approach. If the higher authoritative positions in the hospital know about this Lean amalgamated approach, then the profitability and the quality of the healthcare is balanced very precisely, and the amount of time taken is also minimized. If this method can be used to train employees, qualified and experienced staff will ensure real-worth services for patients as well as the expansion, success, and development of the healthcare organization.

4.5 CONCLUSION

Over the past three decades, the healthcare industry has experienced unprecedented expansion, with rising levels of per capita healthcare spending. A recent epidemic has demonstrated how healthcare facilities around the world can become overburdened by an abrupt increase in the demand for care. To achieve operational excellence and competitive advantage, healthcare businesses must adopt a Lean attitude in conjugation with AI that prioritizes finding timely, high-quality, and cost-effective solutions.

Undoubtedly, the integration of AI and Lean offers tremendous potential in the industry offering healthcare services. The amalgamated approach has the potential to reduce the workloads of employees and improve the quality of the work by improving precision and minimizing error. As a result, fewer unneeded hospital admissions may occur and individuals may be given more control over their health management. Additionally, it might broaden the field of medical understanding and enhance existing clinical advice.

Even if AI and Lean techniques may have advantages in terms of speed and accuracy, medical staff are still needed to perform the activities that need a higher level of cognitive complexity or emotional involvement. The aim behind the change is not to eliminate the existing system, but to concentrate on areas where unneeded resource utilization can be controlled without hampering the quality of care.

4.6 FUTURE SCOPE

The integrated Lean and AI approach is a very promising technology that offers a new practical platform to reduce waste, produce fewer errors, and improve the effectiveness of the healthcare system. The accompanying difficulties, meanwhile, are severe. The process of gathering enough data to train exact algorithms is continuing, necessitating a change in perspective toward data sharing that promotes technical advancement. There is a need for research on the potential and constraints of AI, as well as for clear rules on how to apply and evaluate Lean and AI technology in a safe manner. In the context of Lean techniques, proper training and planning needed to be outlined in order to extract the maximum results from it.

REFERENCES

[1] R. Rathi, A. Vakharia, and M. Shadab, "Lean six sigma in the healthcare sector: A systematic literature review," *Mater. Today Proc.*, vol. 50, pp. 773–781, 2021, doi: 10.1016/j.matpr.2021.05.534

[2] J. Petaschnick, "Health care health care," *Heal. Care Collect.*, vol. 30, no. 5, pp. 10–11, 2016.

[3] D. Tlapa et al., "Effects of Lean healthcare on patient flow: A systematic review," *Value Heal.*, vol. 23, no. 2, pp. 260–273, 2020, doi: 10.1016/j.jval. 2019.11.002

[4] R. Rathi, D. Khanduja, and S. K. Sharma, "A fuzzy MADM approach for project selection: A six sigma case study," *Decis. Sci. Lett.*, vol. 5, no. 2, pp. 255–268, 2016, doi: 10.5267/j.dsl.2015.11.002

[5] Y. Y. M. Aung, D. C. S. Wong, and D. S. W. Ting, "The promise of artificial intelligence: A review of the opportunities and challenges of artificial intelligence in healthcare," *Br. Med. Bull.*, vol. 139, no. 1, pp. 4–15, 2021, doi: 10.1093/bmb/ldab016

[6] L. D. Jones, D. Golan, S. A. Hanna, and M. Ramachandran, "Artificial intelligence, machine learning and the evolution of healthcare: A bright future or cause for concern?," *Bone Jt. Res.*, vol. 7, no. 3, pp. 223–225, 2018, doi: 10.1302/2046-3758.73.BJR-2017-0147.R1

[7] D. S. W. Ting et al., "Development and validation of a deep learning system for diabetic retinopathy and related eye diseases using retinal images from multi-ethnic populations with diabetes," *JAMA – J. Am. Med. Assoc.*, vol. 318, no. 22, pp. 2211–2223, 2017, doi: 10.1001/jama.2017.18152

[8] A. Tufail et al., "Automated diabetic retinopathy image assessment software: Diagnostic accuracy and cost-effectiveness compared with human graders," *Ophthalmology*, vol. 124, no. 3, pp. 343–351, 2017, doi: 10.1016/j.ophtha. 2016.11.014

[9] A. Abdallah, "Implementing quality initiatives in healthcare organizations: Drivers and challenges," *Int. J. Health Care Qual. Assur.*, vol. 27, no. 3, pp. 166–181, 2014, doi: 10.1108/IJHCQA-05-2012-0047

[10] V. Jain and P. Ajmera, "Modelling of the factors affecting Lean implementation in healthcare using structural equation modelling," *Int. J. Syst. Assur. Eng. Manag.*, vol. 10, no. 4, pp. 563–575, 2019, doi: 10.1007/s13198-019-00770-4

[11] M. Alkaabi et al., "Evaluation of system modelling techniques for waste identification in Lean healthcare applications," *Risk Manag. Healthc. Policy*, vol. 13, pp. 3235–3243, 2020, doi: 10.2147/RMHP.S283189

[12] T. Gao and B. Gurd, "Organizational issues for the Lean success in China: Exploring a change strategy for Lean success," *BMC Health Serv. Res.*, vol. 19, no. 1, pp. 1–11, 2019, doi: 10.1186/s12913-019-3907-6

[13] Z. J. Radnor, M. Holweg, and J. Waring, "Lean in healthcare: The unfilled promise?," *Soc. Sci. Med.*, vol. 74, no. 3, pp. 364–371, 2012, doi: 10.1016/j.socscimed.2011.02.011

[14] R. S. Bharsakade, P. Acharya, L. Ganapathy, and M. K. Tiwari, "A Lean approach to healthcare management using multi criteria decision making," *Opsearch*, vol. 58, no. 3, pp. 610–635, 2021, doi: 10.1007/s12597-020-00490-5

[15] H. M. de L. G. Fernandes, M. V. N. de Jesus, D. da Silva, and de E. B. Guirardello, "Lean Healthcare in the institutional, professional, and patient perspective: An

integrative review," *Rev. Gauch. Enferm.*, vol. 41, p. e20190340, 2020, doi: 10.1590/1983-1447.2020.20190340

[16] L. B. M. Costa and M. Godinho Filho, "Lean healthcare: Review, classification and analysis of literature," *Prod. Plan. Control*, vol. 27, no. 10, pp. 823–836, 2016, doi: 10.1080/09537287.2016.1143131

[17] M. Hussain, M. Malik, and H. S. Al Neyadi, "AHP framework to assist Lean deployment in Abu Dhabi public healthcare delivery system," *Bus. Process Manag. J.*, vol. 22, no. 3. 2016, doi: 10.1108/BPMJ-08-2014-0074

[18] C. Jimmerson, D. Weber, and D. K. Sobek, "Reducing waste and errors: Piloting Lean principles at Intermountain Healthcare," *Jt. Comm. J. Qual. Patient Saf.*, vol. 31, no. 5, pp. 249–257, 2005, doi: 10.1016/S1553-7250(05)31032-4

[19] T. Joosten, I. Bongers, and R. Janssen, "Application of Lean thinking to health care: Issues and observations," *Int. J. Qual. Heal. Care*, vol. 21, no. 5, pp. 341–347, 2009, doi: 10.1093/intqhc/mzp036

[20] G. Improta et al., "Lean Six Sigma in healthcare: Fast track surgery for patients undergoing prosthetic hip replacement surgery," *TQM J.*, vol. 31, no. 4, pp. 526–540, 2019, doi: 10.1108/TQM-10-2018-0142

[21] Y. Chen, Y. H. Kuo, P. Fan, and H. Balasubramanian, "Appointment overbooking with different time slot structures," *Comput. Ind. Eng.*, vol. 124, pp. 237–248, 2018, doi: 10.1016/j.cie.2018.07.021

[22] N. Burgess and Z. Radnor, "Evaluating Lean in healthcare," *Int. J. Health Care Qual.Assur.*, vol. 26, no. 3, pp. 220–235, 2013, doi: 10.1108/09526861311311418

[23] P. Ajmera and V. Jain, "A fuzzy interpretive structural modeling approach for evaluating the factors affecting Lean implementation in Indian healthcare industry," *Int. J. Lean Six Sigma*, vol. 11, no. 2, pp. 376–397, 2020, doi: 10.1108/IJLSS-02-2018-0016

[24] V. Jain and P. Ajmera, "Fuzzy TISM and DEMATEL approach to analyse Lean variables in Indian healthcare industry," *Int. J. Process Manag. Benchmarking*, vol. 12, no. 2, p. 233, 2022, doi: 10.1504/ijpmb.2022.121610

[25] A. A. Abdallah, "Healthcare engineering: A Lean management approach," *J. Healthc. Eng.*, vol. 2020, 2020, doi: 10.1155/2020/8875902

[26] P. Achanga, E. Shehab, R. Roy, and G. Nelder, "Critical success factors for Lean implementation within SMEs," *J. Manuf. Technol. Manag.*, vol. 17, no. 4, pp. 460–471, 2006, doi: 10.1108/17410380610662889

[27] J. Harmon et al., "Effects of high-involvement work systems on employee satisfaction and service costs in veterans healthcare," *J. Healthc. Manag.*, vol. 48, no. 6, pp. 393–406, 2003, doi: 10.1097/00115514-200311000-00009

[28] M. Kumar, J. Antony, and A. Douglas, "Does size matter for six sigma implementation? Findings from the survey in UK SMEs," *TQM J.*, vol. 21, no. 6, pp. 623–635, 2009, doi: 10.1108/17542730910995882

[29] R. Patri and M. Suresh, "Factors influencing Lean implementation in healthcare organizations: An ISM approach," *Int. J. Healthc. Manag.*, vol. 11, no. 1, pp. 25–37, 2018, doi: 10.1080/20479700.2017.1300380

[30] M. Enholm, E. Papagiannidis, P. Mikalef, and J. Krogstie, "Artificial intelligence and business value: A literature review," *Inf. Syst. Front.*, pp. 1–26, Aug. 2021, doi: 10.1007/s10796-021-10186-w

[31] A. Bohr and K. Memarzadeh, "The rise of artificial intelligence in healthcare applications," in *Artificial Intelligence in Healthcare*, Elsevier, USA, 2020, pp. 25–60.

[32] "Applications of AI and machine learning in healthcare." https://drkumo.com/ai-and-machine-learning-in-healthcare/ (accessed Oct. 26, 2022).

[33] I. Hazarika, "Artificial intelligence: Opportunities and implications for the health workforce," *Int. Health*, vol. 12, no. 4, pp. 241–245, 2020, doi: 10.1093/INTHEALTH/IHAA007

[34] T. Davenport and R. Kalakota, "The potential for artificial intelligence in healthcare," *Futur. Healthc. J.*, vol. 6, no. 2, pp. 94–98, Jun. 2019, doi: 10.7861/futurehosp.6-2-94

[35] S. Sheikhalishahi, R. Miotto, J. T. Dudley, A. Lavelli, F. Rinaldi, and V. Osmani, "Natural language processing of clinical notes on chronic diseases: Systematic review," *JMIR Med. Inform.*, vol. 7, no. 2. JMIR Publications Inc., Apr. 01, 2019, doi: 10.2196/12239

[36] J. A. Marin-Garcia, P. I. Vidal-Carreras, and J. J. Garcia-Sabater, "The role of value stream mapping in healthcare services: A scoping review," *Int. J. Environ. Res. Public Health*, vol. 18, no. 3. Multidisciplinary Digital Publishing Institute (MDPI), pp. 1–25, Feb. 01, 2021, doi: 10.3390/ijerph18030951

[37] N. A. of E. (US) and I. of M. (US) System, C. on E. and the H. Care, P. P. Reid, W. D. Compton, J. H. Grossman, and G. Fanjiang, "A framework for a systems approach to health care delivery," in *Building a Better Delivery System: A New Engineering/Health Care Partnership*, National Academies Press, USA, 2005, pp. 19–26.

[38] A. Babiker et al., "Health care professional development: Working as a team to improve patient care," *Sudan. J. Paediatr.*, vol. 14, no. 2, pp. 9–16, 2014. [Online]. Available: /pmc/articles/PMC4949805/ (accessed Oct. 26, 2022).

[39] M. E. Kruk et al., "High-quality health systems in the sustainable development goals era: Time for a revolution," *Lancet Glob. Health*, vol. 6, no. 11. Gates Foundation – Open Access, pp. e1196–e1252, Nov. 01, 2018, doi: 10.1016/S2214-109X(18)30386-3

[40] William D. Eggers, David Schatsky, and Peter Viechnicki, "Demystifying artificial intelligence in government | Deloitte Insights," *Deloitte Insights*, 2017. https://www2.deloitte.com/us/en/insights/focus/cognitive-technologies/artificial-intelligence-government.html (accessed Oct. 26, 2022).

[41] Y. Y. M. Aung, D. C. S. Wong, and D. S. W. Ting, "The promise of artificial intelligence: A review of the opportunities and challenges of artificial intelligence in healthcare," *Br. Med. Bull.*, vol. 139, no. 1. Oxford University Press, pp. 4–15, Sep. 01, 2021, doi: 10.1093/bmb/ldab016

[42] A. K. Lawal et al., "Lean management in health care: Definition, concepts, methodology and effects reported (systematic review protocol)," *Syst. Rev.*, vol. 3, no. 1, p. 103, Sep. 2014, doi: 10.1186/2046-4053-3-103

[43] S. T. Teich and F. F. Faddoul, "Lean management – The journey from Toyota to healthcare," *Rambam Maimonides Med. J.*, vol. 4, no. 2, p. e0007, Apr. 2013, doi: 10.5041/rmmj.10107

[44] O. Morell-Santandreu, C. Santandreu-Mascarell, and J. J. Garcia-Sabater, "A model for the implementation of Lean improvements in healthcare environments as applied in a primary care center," *Int. J. Environ. Res. Public Health*, vol. 18, no. 6, pp. 1–33, Mar. 2021, doi: 10.3390/ijerph18062876

[45] D. L. Souza et al., "A systematic review on Lean applications' in emergency departments," *Healthcare (Switzerland)*, vol. 9, no. 6. Multidisciplinary Digital Publishing Institute (MDPI), Jun. 01, 2021, doi: 10.3390/healthcare9060763

[46] N. F. Habidin and S. M. Yusof, "Critical success factors of Lean six sigma for the malaysian automotive industry," *Int. J. Lean Six Sigma*, vol. 4, no. 1, pp. 60–82, 2013, doi: 10.1108/20401461311310526

[47] R. Bercaw, *Lean Leadership for Healthcare*. Productivity Press, New York, 2017.

[48] E. Drotz and B. Poksinska, "Lean in healthcare from employees' perspectives," *J. Heal. Organ. Manag.*, vol. 28, no. 2, pp. 177–195, 2014, doi: 10.1108/JHOM-03-2013-0066

[49] P. Hines, M. Holwe, and N. Rich, "Learning to evolve: A review of contemporary Lean thinking," *Int. J. Oper. Prod. Manag.*, vol. 24, no. 10. Emerald Group Publishing Limited, pp. 994–1011, 2004, doi: 10.1108/01443570410558049

[50] B. Halling and K. Wijk, "Experienced barriers to Lean in Swedish manufacturing and health care," *Int. J. Lean Think.*, vol. 4, no. 2, pp. 43–63, 2013.

Chapter 5

Artificial intelligence as a rescuer of vaccine's cold chain

Palka Mittal
Delhi Pharmaceutical Sciences & Research University, New Delhi, India

Gaurav Aggarwal
Mewat Engineering College, Nuh, India

Puneeta Ajmera
Delhi Pharmaceutical Sciences & Research University, New Delhi, India

Pintu Das
National Institute of Pharmaceutical Education and Research, Guwahati, India

Vineet Jain
Mewat Engineering College, Nuh, India

5.1 INTRODUCTION

The vaccine is a biological product that boosts immunity for a specific disease. It includes typical agents that resemble a disease-causing bacterium and are generated from the microbe's weakened or dead forms, one of its surface proteins, or its toxins. It aids in immune system stimulation, identifies invasive bacteria as foreign invaders, and helps to eradicate them so that the immune system can detect and eradicate any future microorganism it encounters. Vaccines depend on temperature, from the moment of production to the site of administration, and lose their effectiveness over time, when exposed to extreme heat, cold, or light [1, 2]. In 1960, only a few vaccines were available and the first documented success of vaccination was smallpox eradication following the worldwide drive launched by the World Health Organization (WHO) in 1966. Also, the WHO introduced the Expanded Programme on Immunization in 1974 [3]. Immunization has also played an essential role in the fight to eradicate polio [4]. On a global scale, the Expanded Programs on Immunization (EPI) depend on effective vaccine supply chains to provide children with effective and potent vaccines in an equitable, timely, and effective way [5].

DOI: 10.1201/9781003462163-5

Incredible strides have been made in the direction of achieving universal vaccination coverage in the 40 years since the introduction of the EPI. This development is the result of numerous distinct stakeholders, policies, technology, and initiatives. However, the most fundamental component of this system is the infrastructure of vaccine supply chains. This infrastructure covers everything from the machinery that keeps vaccines cold to the logistics and information systems that monitor and control supply flows to the planning of the country's distribution systems and the staff members who oversee them [6].

Unfortunately, one of the prevalent obstacles preventing equal and full access to the advantages of vaccination is a lack of adequate immunization supply chain and logistics structures. These gaps affect the accessibility and efficacy of vaccinations at the site of administration, hinder the development of novel, life-saving vaccines, and also waste valuable human and economic resources [7]. Every year, pharmaceutical companies continue to lose millions of dollars due to these instances [8].

The current advancement of technology, Artificial Intelligence (AI), the Internet of Things (IoT), cloud computing, big data, and other communication technologies, as well as the rapid integration with other cutting-edge technology, has facilitated the expansion of cold chain logistics. In the cold chain logistics sector, "big data" has become a buzzword, and the importance of studying huge data has become increasingly apparent, especially when big data and IoT are combined. The operational effectiveness of cold chain logistics can be increased with the acquisition of big data, lower energy usage, and improved resource allocation, which is in line with the development of the present-day green ecological economy.

5.2 THE VACCINE COLD CHAIN AND ITS CHALLENGES

Immunization is largely regarded as the most economical and effective public health treatment in history, having been estimated to save almost two million lives per year [9]. Every year, immunization prevents two to three million deaths and lengthens life expectancies [4]. On the other side, 19.5 million infants globally still lack access to essential vaccinations. The 10 nations, namely, Afghanistan, Angola, the Democratic Republic of Congo, Ethiopia, India, Indonesia, Iraq, Nigeria, Pakistan, and South Africa are home to almost 60% of these children. Other developing nations experience a remarkably similar predicament. To ensure that no infant dies from an illness that can be prevented by vaccination, the government intends to immunize every newborn infant. A well-designed, effective, and efficient vaccine supply chain/cold chain system can help to fulfill the objective of guaranteeing that every child is vaccinated [10].

Cold chain refers to maintaining the appropriate temperature across all of the steps through which vaccine transit, from their manufacturing, transportation, storage, delivery, sale, and lastly consumer [11, 12].

5.2.1 Components of the vaccine cold chain

A vaccine cold chain is primarily concentrated on cold chain infrastructures, which include temperature regulation, temperature monitoring, temperature indication, and cold chain management systems. The critical components of the vaccine supply chain are production, storage, shipping, and administration. The cold chain for vaccines primarily focuses on how vaccines are transported and stored [13]. The components of the chain are shown in Figure 5.1.

(a) **Product:** Policymakers must choose which disease to target and which vaccine to deploy before immunization can begin. It's possible that different vaccines are available for the same disease or that the disease's features were unknown while the vaccine was being created. This raises the issue of choosing the vaccine's composition. Designing a vaccination program for numerous diseases requires making choices about which vaccinations to utilize. These involve policymaking issues

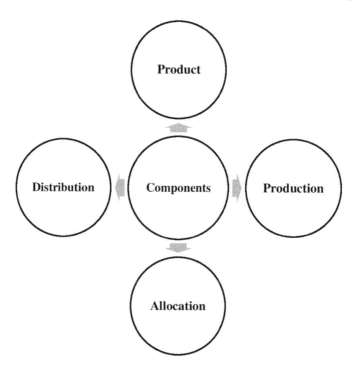

Figure 5.1 Major components of the vaccine supply chain.

such as: which diseases to include, which vaccines to employ, and how vaccinations should be planned in the program.

(b) **Production:** Vaccination production is characterized by yield and lead time uncertainty, which may lead to inefficiencies in the vaccine market. The alignment of supply and demand can be improved by market coordination.

(c) **Allocation:** Especially during rapid outbreaks, the amount of vaccine doses available is frequently insufficient to vaccinate the entire population. It generates a need to identify high-risk and low-risk people as well as high- and low-transmission populations within a community. Also, when an outbreak spreads across borders, (re)allocation issues between various regions and/or nations may occur.

(d) **Distribution:** Distributing the vaccinations from the manufacturer to the ultimate consumers is the final phase. Making judgments on vaccine stockpile locations involves inventory control. The location, staffing levels, and design of fixed distribution sites are all logistical considerations [14].

To deliver vaccines to children, the vaccine supply chain (VSC) adheres to the "six rights", i.e., right vaccine, right quantity, right time, right place, right condition, and right cost. Research has also addressed the subject of how poorly constructed cold chains and unanswered problems in the supply chain of essential vaccines contribute to low child immunization coverage. The smooth operation of immunization programs can be disrupted by supply chain problems that may develop at any point in the vaccine distribution process [10]. The biggest challenges a cold chain faces are power outages or the lack of backup systems to maintain the required temperature, the lack of real-time alerts for temperature deviations, unknown causes of temperature deviations, locating the cold chain when it is traveling, vaccine deterioration while waiting to be transferred from a storage room to a carrier, and creating a report for the entire chain from beginning to end.

5.2.2 Challenges to vaccine cold chain

The vaccine cold chain is currently under the burden of several barriers that impact its successful execution and hence hinder immunization coverage. These are outlined in Table 5.1.

Among all of these, temperature sensitivity is the major concern while dealing with vaccines. Temperature excursion is defined as an occurrence in which a time-temperature-sensitive pharmaceutical product is exposed to temperatures outside of the acceptable ranges [12]. Each vaccine has a different level of sensitivity to heat, ice, and light. According to each vaccine's stability properties, they must be maintained within a specific temperature range and must be shielded from extremes of heat and cold [4]. Heat exposure can

Table 5.1 Challenges to the vaccine supply chain

Problem	Scale of problem	Cause of problem	Consequences
Inadequate capacity of the cold chain [5, 9]	The increased need for storage space is largely driven by the launch of new vaccines rather than routine and population growth.	• Lack of awareness of or comprehension of the status of the cold chain equipment and the status of inventory. • Lacking in forecasting upcoming capacity needs • Insufficient systems for implementation of plans to close existing gaps • Insufficient finances • Improper management and monitoring	• The introduction of new vaccinations may get postponed • Delivery costs rise as delivery frequency is increased • If vaccinations are moved to locations based on capacity instead of need, vaccine availability and coverage may get affected • A malfunction of the system can cause damage to the vaccine. • The unequipped facility may struggle to ensure the safety of vaccines
The absence of "optimal" or cutting-edge technology [9]	Outdated equipment is unable to offer the protective benefit of contemporary designs, due to ineffective temperature control, lower stay times, and no freeze protection.	• Unawareness of the benefits of updated technology • Difference in cost • High cost of the equipment with the latest technology	• Temperature damage to vaccines • Difficult to identify the right time for equipment maintenance
Poor temperature monitoring and upkeep system [5, 3, 9, 15]	Functional difficulties and serious temperature issues are most common and frequently occur at the facility level.	• Unawareness of the impact of fluctuation in temperature • Absence of real-time temperature monitoring • Unavailability of proper resources	• Loss and wastage of deteriorated vaccines due to uncontrolled temperature.
Lack of well-qualified staff [3, 9]	Lack of mid-level managers or logisticians, and no unified policies, practices, or training resources that could be used as guidelines.	• Absence of policy and guidelines for proper training	• Lack of monitoring • Risk of vaccine damage as unqualified staff dealing with them.

impair vaccines by changing the tertiary protein structure, detaching the polysaccharides from the protein carriers in polysaccharide conjugate vaccines, and reducing the pathogenicity of live-attenuated vaccines. The cold chain would be put under additional strain if vaccines were frozen because this could result in a clustering of the adjuvanted particles, antigen degradation, and molecular alterations [13]. The need for proper temperatures and conditions for vaccines to maintain their efficacy is well established, but the difficulties, dangers, and flaws in the accompanying cold chain that allows vaccines to leave producers, travel across the globe, and eventually reach a patient's arm are less well understood [4].

Systems for managing the cold chain and logistics of vaccines are at the heart of the biggest obstacles to ensuring the highest vaccination coverage. Although several new vaccines have been distributed throughout the developing world (and more vaccines are in development), cold chain systems are still experiencing difficulties. Challenges in the vaccine supply chain may result in:

(a) administered vaccines not being as effective as intended (for example, because of inadequate temperature control or non-operational equipment)
(b) issues with the unavailability of vaccines (due to insufficient storage capacity, interrupted service delivery, vaccine shortage, etc.)
(c) the inefficient use of available material labor and finances (such as losses due to vaccine waste) [9].

To overcome the existing challenges and to prevent upcoming issues NGCA (New generation of the supply chain for health products) has demanded a vaccine supply chain with the following characteristics:

(a) A supply chain that is simplified and effective, with less waste, and reduced logistics expenses.
(b) The implementation of proper information technology solutions, to ensure clear data visibility of vaccination availability and quality to the point of delivery.
(c) Modern cold chain equipment with regular maintenance and monitoring.
(d) A specialized logistics workforce with the power to decide on the supply chain based on information and real-world situations.
(e) The involvement of key performance indicators (KPIs) to assess system performance, encourage evidence-based decision-making, and direct continuous development.
(f) Consistent finance and financial flows across the whole supply chain to enable regular, dependable commodity deliveries [16].

5.3 ARTIFICIAL INTELLIGENCE IN THE IMMUNIZATION SUPPLY CHAIN

Temperature is a numerical value that expresses how much heat a product contains. High (hot) and low (cold) temperature are concepts that are related to this magnitude. The temperature-measuring equipment must be accurate to about 0.5°C and have a range of −30 to 20°C [11]. To maintain temperature control and maintenance throughout the process, AI technology is crucial to the cold chain.

Studies have examined the effects of AI and its potential to substitute humans with intelligent automation that benefits production, development, and even distribution [17, 18]. In comparison to manual procedures, AI-centric solutions are capable of monitoring and managing operations in real time [19, 20].

5.3.1 Applications of AI in SCM and logistics

(a) **Procurement:** The goals of procurement through AI are to support operational needs, manage to purchase effectively, develop, select, and maintain supply sources, support organizational goals and objectives, in forging strong bonds with other units and organizations, and in creating purchase strategies that support the maintenance of organizational strategies.

(b) **Manufacturing:** AI-based solutions facilitate the rapid data-driven decision-making process and the digital transformation of manual product manufacturing processes.

(c) **Warehousing:** The Warehouse Management System (WMS) controls the inventory control of items received from suppliers and vendors and held in warehouses. They are employed to locate the inventory in the manufacturing, processing, and production facilities and manage customer orders accordingly.

(d) **Packaging:** AI is used in packaging to handle and distribute products, adjust product density, recognize and identify products, and enhance productivity. With the aid of barcodes and electronic data transmission, it facilitates straightforward return management (EDI).

(e) **Distribution:** AI facilitates and speeds up the physical distribution process as well as communication and facilitation tasks. Tools gather the necessary data and make recommendations for actions to streamline processes and increase earnings. AI assists in anticipating consumer needs so that the clients are effectively and efficiently served with this knowledge.

(f) **Customer relation management:** AI provides collaborative planning, forecasting, and replenishment to manage vendor inventory Collaborative Planning, Forecasting and Replenishment (CPFR). It assists in continuously learning about clients, addressing them differently, and anticipating their needs [21].

5.3.2 Opportunities to AI

(a) Understanding the opportunities (consumer demands, attitudes, and preferences), and risks (degradation, wastage, and shortage) will help in taking the necessary countermeasures.
(b) Analysis and interpretation of data to understand and manage issues and failure, in order to increase efficiency.
(c) Interacting with people (including staff, service users, and customers) by utilizing natural language processing and other sensory skills.
(d) Automating inventory management by knowing the market demand and the involvement of advanced intervention.
(e) Without valid and updated intervention, humans are "radically unresponsive to both the quality and amount of information that gives rise to impressions and intuitions", which can be combated by the application of AI.
(f) Assisting in enhancing the reaction to adapt to environmental changes [22].

5.3.3 Advantages of AI

(a) AI is included in the information system, which keeps the information flow rich.
(b) Real-time monitoring is one of its most common applications.
(c) AI techniques are used to construct profiles of diverse in-sources and out-sources.
(d) Systems based on rules could aid in contract decision-making and the outsourcing of logistics.
(e) The location might be investigated, and AI-based machine learning could be used to solve mid-way problems.

5.3.4 Disadvantages of AI

(a) Because AI solutions are complex and hard to understand, they are tough to implement.
(b) Since AI solely relies on the program, it occasionally makes mistakes.
(c) Due to information acquisition bottlenecks, AI cannot handle the risks and uncertainties associated with its cross-functional nature.

5.3.5 Challenges to AI

With the advancements and numerous benefits, AI also has multiple challenges that are shown in Figure 5.2.

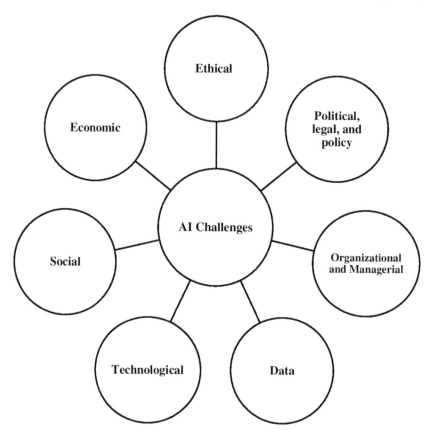

Figure 5.2 Challenges to AI.

(a) **Social:** One of the primary obstacles to the adoption of AI is culture, as patients may be hesitant to engage with novel devices and
programs. It also emphasized the need for patient knowledge, education, and awareness to allay the fear of advanced technology among
many patient groups and for doctors to understand how to deal
with implementation and execution of AI technologies healthcare
infrastructure.

(b) **Economic:** AI is an expensive technology and hence it increases the
cost of computations, products/services, and patient treatment.

(c) **Ethical:** Medical professionals and patients may display a lack of trust
and concerns over the ethical implications of AI systems and how they
use shared data. Due to the rapid pace of modification and advancement of AI technology, there are increasing worries that ethical issues
are not being properly addressed.

(d) **Political, legal, and policy**: It includes difficulties with copyright, autonomous intelligence system governance, accountability and responsibility, privacy/safety, dangers to national security posed by foreign-owned firms gathering personal information, lack of accountability standards while using AI, and absence of recognized industry standards for the application and assessment of AI.

(e) **Organization and managerial**: Lack of knowledge of AI and related development strategies is a challenge at the organizational level. It also generates a threat of unemployment by replacing the human workforce.

(f) **Data**: An absence of data to support the advantages of AI solutions, dimensional challenges, repeatability and transparency issues, data quantity and quality, and lack of data-collecting standards are a few concerns related to data while implementing AI.

(g) **Technological**: Studies have examined the difficulties of using AI technology for data and image interpretation as well as the non-Boolean character of diagnostic jobs in healthcare. Among the prevalent technical obstacles are architecture problems, difficulties processing unstructured data, and adversarial assaults [22].

5.4 DIFFERENT TECHNOLOGIES IN THE COLD CHAIN

The last ten years have seen the development of several successful works that employ various technologies in cold chains. Among them are Wireless Sensor Networks (WSN), IoT, Radio Frequency Identification (RFID) sensors, and GPRS (General Packet Radio Service) labels. The WSN, also known as sensor and actuator networks (WSAN), is a group of spatially dispersed autonomous sensors used to track various physical and environmental parameters, including temperature and humidity [11]. AI and IoT are revolutionizing several industries with their powerful monitoring and quick decision-making capabilities [23]. A smart traceability system that incorporates wireless sensor networks and statistical process management for the cold chain is suggested [24].

5.4.1 Internet of Things (IoT)

A global organized network called the Internet of Things (IoT) connects common things with intelligence, identification, and sensing capabilities. The idea of the IoT was developed by Kevin Ashton, a British entrepreneur. It was created in 1999 and encapsulates the concept of the physical world connecting with computers through ubiquitous sensors [25]. Smart devices that perform activities and communicate real-time data are referred to as IoT when they are connected to sensors and the Internet [26].

The usage of the IoT, RFID and GPRS technologies is suggested for a real-time temperature monitoring system. The idea, which was not developed in the other phases of the chain, concentrates on the chilled transportation of goods [27]. It is suggested that a system of WSN integrated with IoT can be used to create an intelligent tracking system for the supply chain. A microprocessor is used by the system to control operations [28]. WSN and statistical quality control based on IoT architecture are used to provide a cold chain structure with a time-temperature indicator (TTI) [29]. It is suggested that we should use an IoT-based load monitoring system (IoT-CMS) to keep track of any changes in environmentally sensitive items and ensure their operational condition throughout the cold chain in the working atmosphere [30]. The use of these devices was previously unthinkable, mostly because of their high cost, but stakeholders are only now initiating to incorporate these technology into the cold chain, and it is projected that this incorporation will continue in the upcoming time [31].

Pharmaceutical supply chain management is well adapted to IoT applications such as cold chain monitoring, resource (man and equipment) tracking, packaging, and warehousing management. Pharmaceutical companies have a compelling potential to adopt and profit from the IoT, the game-changing technology, since it can assist manage supply chains, improving services, and creating products [32]. Transparent cold chain monitoring is achieved by using the IoT paradigm to track and trace products and environmental factors automatically along the cold chain. Additionally, real-time measurements of the quality decline are possible [33]. IoTs are utilized in supply chain management in the areas of manufacturing, logistics, quality control, inventory management, and real-time decision-making [34, 35].

For cold chain operators, the integration of IoT, WSN, and RFID technology has many benefits [11, 36], as follows:

1) They make it possible to keep an eye on the cold chain's characteristics.
2) They lower gasoline expenses by keeping an eye on traffic in real time.
3) By identifying temperature differences and taking preventive action, you can help to lower food loss.
4) They make it simpler to control the inventory in the warehouse.
5) They enable notifications to be sent to cloud-connected devices.
6) They automatically keep track of the temperature change in a product.
7) It offers an early warning system when the medications at a specific cold chain link exceed the standard.

5.4.1.1 Benefits of implementing IoT in the cold chain

1. IoT provides solutions that were not previously available to tackle cold chain issues, particularly with regard to increasing cold chain responsiveness.

2. It can be used to improve the responsiveness of the cold chain and to find whether real-time monitoring is performed.

3. It is beneficial for the cold chain of vaccines by taking into account the main characteristics of the cold chain, such as range of temperature, and controlled temperature [36].

5.4.1.2 IoT-based conceptual model

The development of a conceptual model for the monitoring of temperature throughout the entire vaccine supply chain depends on having a thorough understanding of the fundamental traits and difficulties faced by the pharmaceutical and healthcare industries.

(a) **Need to measure temperature:** To keep the product's specified quality up until it reaches the final consumer, temperature control at every point in the supply chain is crucial [37].

A thorough logistical strategy must be established to move products via the cold chain without experiencing temperature fluctuations. This process includes shipment preparation, transportation, customs procedures, distribution, and final integrity certification at the point of delivery [36].

(b) **Risk assessment:** Each node in the cold chain system is connected with the one before it and also influences the one after it. A chain connection failure invariably causes a loss in quality and revenue [38]. Risk analysis is essential, and it necessitates ongoing product control. To maintain both the quality and quantity of a product, visibility is still crucial [37]. Another important component is the disruption period or the time from the incident or interruption that affects the distribution indication until the supply chain system stabilizes and can start performing its role.

(c) **Complexity:** Through complicated chains, there is a greater chance of disturbance. The pharmaceutical industry's supply chain is disorganized and very complex. Every nation on earth is a part of this network, which further extends this intricacy on a global scale [39].

(d) **Leadership:** The healthcare sector is still fragmented and lacks true leadership despite all of the consolidation which has occurred. Fragmentation makes it more challenging to connect the hundreds of stakeholders participating at each stage and to ensure effective communication between them [36].

(e) **Product Recall:** Dissimilar to other products, pharmaceutical products are destroyed when they are recalled or returned. As a result, there are several factors to take into account while considering reverse logistics: the necessity for precise tracking and visibility, requirements for the cold chain, appropriate storage, and anti-counterfeiting [36].

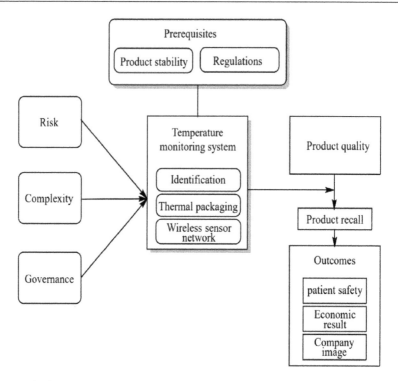

Figure 5.3 Conceptual model.

The WSN with RFID tags incorporated can enhance chain visibility, which is crucial to maintain both the quality and the quantity of any product. The relationship between product recall and results is also indirectly moderated is shown in Figure 5.3 [36].

5.4.2 Big data

The term "big data" has gained popularity in the vaccine supply chain logistics sector, and research into big data has proved enormously significant, particularly when big data is combined with the IoT. The growth of the modern green ecological economy is in line with the extraction and collection of data in cold chain logistics, which can enhance operational effectiveness, optimize resource usage, and lower consumption of energy.

Modern information technology has advanced quickly, particularly with the widespread application and embedding of data, AI, IoT, and other technologies. As a result, the logistics operations are currently displayed in an associative, connected, and intelligent manner, enabling real-time cold chain logistics monitoring [40].

5.4.2.1 Encourage real-time dynamic product information feedback

By placing a temperature and humidity sensor in the carriers, communicating the information to the remote areas, and combining it with historic data, the logistics business may determine the best conditions for holding the vaccines. This will help to reduce product loss [40].

5.4.2.2 Encourage the improvement and modification of the application approach

The system may combine real-time data with data interpretation and analysis techniques to produce useful and rapid results, emphasizing the relevance of accurate data in cold chain logistics. Timeliness is also the most crucial element in the logistics of the cold chain. Due to the more unique nature of the items, there may be considerable seasonal variations, raising the cost and risk for logistics businesses to some level. Big data technology can be used to optimize and make necessary adjustments to the internal and external elements and marketing strategies with which logistics companies must deal [40].

5.4.3 Blockchain

In recent years, Blockchain has developed into a promising resource that has the potential to be simple and effective in addressing a wider range of difficulties, particularly in this age of digitization. Blockchain in integration with existing technologies can prove helpful in preventing the efficacy of vaccines. AI, cloud computing, machine learning, and IoT are just a few of the existing technologies that complement Blockchain nicely. Such convergences are even more encouraging in terms of managing the cold chain for vaccines, particularly in terms of ensuring and later monitoring the efficacy of vaccines at the time of delivery to the end beneficiaries. The obvious truth is that Blockchain is very capable in the sense that it's the most ethical means of gathering data. This method was utilized to gather the information that was fetched from the units about its coverage, and, as a result, the inaccessible demographics were known [41].

5.4.4 Cloud computing

Technology change has been significantly impacted by cloud computing. Through efficiently lowering the expenses of information technology software, this can supply essential software for corporate management. Based on cloud computing, a cold chain logistics system is created. This technology improves communication between cold chain logistics and its clients, creates a shared database of products, speeds up distribution, and maximizes interests [42].

5.4.5 Machine learning

AI is a set of ideas, techniques, and technologies that aims to better understand intelligence and provide machines with abilities that would be considered intelligent when used by people or other living things. The area of AI known as machine learning examines these methods and algorithms [31].

Examples of how machine learning has been used in studies of the cold chain are included in Table 5.2 [31].

5.4.6 Artificial neural network (ANN)

Artificial neural network (ANN) has been a component of AI since the 1950s when it first began to take shape. However, a revolutionary leap was made in 1985 with the creation of the back-propagation algorithm for multilayer perceptions, which enabled neural networks to begin completing significant learning tasks, particularly from raw data, which were previously unachievable by the majority of the learning techniques that were in use at the time. Over the past ten years, interest in ANN has grown significantly because of the combination of huge training databases, a major rise in processing power, and new technical features [31].

ANN, or connectionist systems, are based loosely on the biological neural networks that make up animal brains. For modeling, structuring, and research objectives, the ANN has been used. The ANN technique presupposes that the source of temperature and the goal temperature (the temperature inside a pallet) have a nonlinear relationship [11]. The data related to temperature gathered for the management of cold chain is tracked using ANN and exponentially weighted moving average control (EWMA). The work's main objective is to statistically control the temperature data that was gathered. The accuracy of the temperature model used does not promise that

Table 5.2 Examples of cold chain investigations utilizing machine learning

Models	Target	References
Back Propagation Neural Network	Predict variations and fluctuations in temperature	[38]
Neural Network	Prediction of temperature in the pallet	[43]
Neural Network	Estimation of temperature across pallet	[44]
Firefly algorithm	Reduce risk automatically	[45]
Neural Network	Prediction of temperature inside pallet	[46]
Kernel Logistic Regression	Support in decision-making to ensure the quality	[47]
Neural Network	Prediction of temperature	[48]
Compressed Sensing	CO_2 signal reconstruction	[49]

it can be used for process monitoring [38]. The synchronous temperature monitoring in the cold chain is improved with the help of a better quality of detectability (QOD) algorithm [50].

5.5 CONCLUSION

In the past five years, the phrase "supply chain" has taken the place of the "vaccine cold chain." The name denotes an evolution of the exclusive distribution and storage of vaccinations to a strategy that includes both vaccines and medications. As the quantity of vaccines continues to rise, the range of vaccination operations broadens, and new target populations are discovered, the supply chain's design needs to be simplified and enhanced. Other factors influencing change in the immunization supply chain include the rising use of information and communications technology, particularly mobile phones; improved access to data about vaccines and immunization; the need for significantly more storage space to accommodate the increasing number of vaccines; and the significantly higher costs for some newer vaccines, necessitating a decrease in vaccine wastage rates [3].

When handling sensitive products in supply chains, hazard management is essential for lowering the loss of product and resources. On the one hand, the products are more likely to deteriorate or become contaminated during delivery in the cold chain because of the alterations in temperature and humidity. The inefficient capture of visible product monitoring data may suffer needless loss. On the other hand, the majority of operations and procedures need a lot of labor, and the operators must operate in challenging conditions.

AI in integration with other technological advancements has made it possible to achieve the "ideal path for dispatching cold chain logistics and distribution", "promoting real-time dynamic feedback of product information", "building a sound standardization system", and "promoting standardization of product production and circulation". Single temperature monitoring has given way to multiple parameter monitoring, basic information analysis give way to integrated system modeling, and manual management gives way to intelligent management in the vaccine supply chain. It ensures that the entire distribution of vaccines from the point of production to the client always takes place in the regulated environment needed to sustain quality while also ensuring the most effective use of energy along the cold chain.

5.6 FUTURE SCOPE

Vaccines provide a high return on investment and can significantly reduce the prevalence of morbidity and mortality from several major diseases. Therefore, improvement and modifications in vaccine supply chain systems can increase effective immunization coverage and lower the number of deaths brought by diseases that can be prevented by vaccination.

Significant and consistent performance improvement requires moving from problem-solving to a continuous improvement process of identifying, mitigating, and preventing underlying problems.

Although the combination of the cold chain and AI have many benefits, there are still some difficulties. To tackle these, there is a requirement for more sophisticated and powerful AI technologies. The interaction between energy usage and the environment, as well as the relationship between the quality and efficiency of vaccination products, must also be balanced while optimizing cold chain management solutions. To do this, AI must seek the best possible outcomes for several competing goals, including product quality, energy use, and efficiency. AI and advanced technology, used in conjunction with low-power sensors, cloud computing, and the IoT, will gain ground.

REFERENCES

1. Hanson CM, George AM, Sawadogo A, Schreiber B. Is freezing in the vaccine cold chain an ongoing issue? A literature review. *Vaccine* [Internet]. 2017;35(17):2127–2133. Available from: http://dx.doi.org/10.1016/j.vaccine.2016.09.070

2. Yakum MN, Ateudjieu J, Pélagie FR, Walter EA, Watcho P. Factors associated with the exposure of vaccines to adverse temperature conditions: The case of North West region, Cameroon Health Services Research. *BMC Res Notes*. 2015;8(1):1–7.

3. Lloyd J, Cheyne J. The origins of the vaccine cold chain and a glimpse of the future. *Vaccine* [Internet]. 2017;35(17):2115–2120. Available from: http://dx.doi.org/10.1016/j.vaccine.2016.11.097

4. Kartoglu U, Ames H. Ensuring quality and integrity of vaccines throughout the cold chain: The role of temperature monitoring. *Expert Rev Vaccines* [Internet]. 2022;21(6):799–810. Available from: https://doi.org/10.1080/14760584.2022.2061462

5. Srivastava V, Ratna M, Ray A, Shukla S, Shrivastava V, Kothari N, et al. Strengthening the immunization supply chain: A time-to-supply based approach to cold chain network optimization & extension in Madhya Pradesh. *Vaccine* [Internet]. 2021;39(45):6660–6670. Available from: https://doi.org/10.1016/j.vaccine.2021.09.062

6. Brooks A, Habimana D, Huckerby G. Making the leap into the next generation: A commentary on how Gavi, the Vaccine Alliance is supporting countries' supply chain transformations in 2016–2020. *Vaccine* [Internet]. 2017;35(17):2110–2114. Available from: http://dx.doi.org/10.1016/j.vaccine.2016.12.072

7. Brison M, LeTallec Y. Transforming cold chain performance and management in lower-income countries. *Vaccine*. 2017;35(17):2107–2109.

8. Sharma R, Kamble SS, Gunasekaran A, Kumar V, Kumar A. A systematic literature review on machine learning applications for sustainable agriculture supply chain performance. *Comput Oper Res* [Internet]. 2020;119:104926. Available from: https://doi.org/10.1016/j.cor.2020.104926

9. Ashok A, Brison M, LeTallec Y. Improving cold chain systems: Challenges and solutions. *Vaccine* [Internet]. 2017;35(17):2217–2223. Available from: http://dx.doi.org/10.1016/j.vaccine.2016.08.045

10. Chandra D, Kumar D. Prioritizing the vaccine supply chain issues of developing countries using an integrated ISM-fuzzy ANP framework. *J Model Manag.* 2020;15(1):112–165.

11. Campos Y, Villa JL. Technologies applied in the monitoring and control of the temperature in the cold chain. *2018 IEEE 2nd Colomb Conf Robot Autom CCRA 2018.* 2018;1–6.

12. Hernández A, Yamaura M. Cold chain management with Internet of Things (IoT) enabled solutions for pharmaceutical industry. *MisiEduMy* [Internet]. 2022;(2011):1–6. Available from: https://www.misi.edu.my/wp-content/uploads/thesisresearch/2018/Atziri-Momoru.pdf

13. Business G, Research M. Characteristics, challenges, and opportunities of vaccine cold chain. *An Int J.* 2021;13(3).

14. Duijzer LE, van Jaarsveld W, Dekker R. Literature review: The vaccine supply chain. *Eur J Oper Res* [Internet]. 2018;268(1):174–192. Available from: https://doi.org/10.1016/j.ejor.2018.01.015

15. Kristensen DD, Lorenson T, Bartholomew K, Villadiego S. Can thermostable vaccines help address cold-chain challenges? Results from stakeholder interviews in six low- and middle-income countries. *Vaccine* [Internet]. 2016;34(7):899–904. Available from: http://dx.doi.org/10.1016/j.vaccine.2016.01.001

16. Numbi FKM, Kupa BM. The potential of next-generation supply chains to ease DRC's "Casse-tête". *Vaccine.* 2017;35(17):2105–2106.

17. Kusiak A. Artificial intelligence and operations research in flexible manufacturing systems. *INFOR Inf Syst Oper Res.* 1987;25(1):2–12.

18. Muhuri PK, Shukla AK, Abraham A. Industry 4.0: A bibliometric analysis and detailed overview. *Eng Appl Artif Intell* [Internet]. 2019;78(September 2018):218–235. Available from: https://doi.org/10.1016/j.engappai.2018.11.007

19. Jain PK, Mosier CT. Artificial intelligence in flexible manufacturing systems. *Int J Comput Integr Manuf.* 1992;5(6):378–384.

20. Zhong RY, Xu X, Klotz E, Newman ST. Intelligent manufacturing in the context of industry 4.0: A review. *Engineering* [Internet]. 2017;3(5):616–630. Available from: http://dx.doi.org/10.1016/J.ENG.2017.05.015

21. Niranjan K, Narayana KS, Rao MVALN. Role of artifical intelligence in logistics and supply chain. *2021 Int Conf Comput Commun Informatics*, ICCCI 2021. 2021;27–29.

22. Dwivedi YK, Hughes L, Ismagilova E, Aarts G, Coombs C, Crick T, et al. Artificial Intelligence (AI): Multidisciplinary perspectives on emerging challenges, opportunities, and agenda for research, practice and policy. *Int J Inf Manag.* 2021;57(July):101994.

23. Zhang Z, Liu XY, Zhang XR. The role of artificial intelligence in energy aspects of super cold chain of agricultural products. *Int J Energy Res.* 2022;46(April):1–6.

24. Xiao X, Fu Z, Zhang Y, Peng Z, Zhang X. Developing an intelligent traceability system for aquatic products in cold chain logistics integrated WSN with SPC. *J Food Process Preserv.* 2016;40(6):1448–1458.

25. Witkowski K. Internet of Things, big data, industry 4.0 – Innovative solutions in logistics and supply chains management. *Procedia Eng* [Internet].

2017;182:763–769. Available from: http://dx.doi.org/10.1016/j.proeng.2017.03.197

26. Radoglou Grammatikis PI, Sarigiannidis PG, Moscholios ID. Securing the Internet of Things: Challenges, threats and solutions. *Internet of Things (Netherlands).* 2019;5:41–70.

27. Wu LZ, Zhao Y. Cold chain logistics temperature monitoring system based on internet of things technology. *Appl Mech Mater.* 2013;416–417:1969–1973.

28. Luo H, Zhu M, Ye S, Hou H, Chen Y, Bulysheva L. An intelligent tracking system based on internet of things for the cold chain. *Internet Res.* 2016; 26(2):435–445.

29. Shih CW, Wang CH. Integrating wireless sensor networks with statistical quality control to develop a cold chain system in food industries. *Comput Stand Interfaces* [Internet]. 2016;45:62–78. Available from: http://dx.doi.org/10.1016/j.csi.2015.12.004

30. Tsang YP, Choy KL, Wu CH, Ho GTS, Lam HY, Koo PS. An IoT-based cargo monitoring system for enhancing operational effectiveness under a cold chain environment. *Int J Eng Bus Manag.* 2017;9:1–13.

31. Loisel J, Duret S, Cornuéjols A, Cagnon D, Tardet M, Derens-Bertheau E, et al. Cold chain break detection and analysis: Can machine learning help? *Trends Food Sci Technol.* 2021;112(April):391–399.

32. Sharma DK, Gupta P. Internet of Things: The new Rx for pharmaceutical manufacturing and supply chains. *An Industrial IoT Approach for Pharmaceutical Industry Growth.* 2020;2:257–288.

33. Tsang YP, Choy KL, Wu CH, Ho GTS, Lam CHY, Koo PS. An Internet of Things (IoT)-based risk monitoring system for managing cold supply chain risks. *Ind Manag Data Syst.* 2018;118(7):1432–1462.

34. Goodarzian F, Navaei A, Ehsani B, Ghasemi P, Muñuzuri J. Designing an integrated responsive-green-cold vaccine supply chain network using Internet-of-Things: Artificial intelligence-based solutions. *Ann Oper Res* [Internet]. 2022. Available from: https://doi.org/10.1007/s10479-022-04713-4

35. Ben-Daya M, Hassini E, Bahroun Z. Internet of things and supply chain management: A literature review. *Int J Prod Res* [Internet]. 2019;57(15–16):4719–4742. Available from: https://doi.org/10.1080/00207543.2017.1402140

36. Monteleone S, Sampaio M, Maia RF. A novel deployment of smart cold chain system using 2G-RFID-Sys temperature monitoring in medicine cold chain based on Internet of Things. *Proc – 2017 IEEE Int Conf Serv Oper Logist Informatics*, SOLI 2017. 2017; 2017-Janua:205–210.

37. Bogataj M. Stability of perishable goods in cold logistic chains. *Int J Prod Econ.* 2005;93–94(SPEC.ISS.): 345–356.

38. Chen KY, Shaw YC. Applying back propagation network to cold chain temperature monitoring. *Adv Eng Informatics* [Internet]. 2011;25(1):11–22. Available from: http://dx.doi.org/10.1016/j.aei.2010.05.003

39. Rossetti CL, Handfield R, Dooley KJ. Forces, trends, and decisions in pharmaceutical supply chain management. *Int J Phys Distrib Logist Manag.* 2011;41(6):601–622.

40. Sun X, Gao L, Liang Y. Research on big data acquisition and application of cold chain logistics based on artificial intelligence and rnergy Internet of Things. *IOP Conf Ser Earth Environ Sci.* 2019;252(5):1–7.

41. Vargis BK, Alam MS, Upreti K, Raut RD, Kumar N, Jalil SZA. Scoping analysis of leveraging IoT with blockchain for monitoring and ensuring efficacy of vaccine cold chains. *Proc – 2022 6th Int Conf Intell Comput Control Syst ICICCS 2022.* 2022;(ICICCS): 795–803.

42. Adhianto L, Banerjee S, Fagan M, Krentel M, Marin G, Mellor-Crummey J, et al. HPCTOOLKIT: Tools for performance analysis of optimized parallel programs. *Concurr Comput Pract Exp.* 2010;22(6):685–701.

43. Mercier S, Uysal I. Neural network models for predicting perishable food temperatures along the supply chain. *Biosyst Eng* [Internet]. 2018;171:91–100. Available from: https://doi.org/10.1016/j.biosystemseng.2018.04.016

44. Badia-Melis R, Qian JP, Fan BL, Hoyos-Echevarria P, Ruiz-García L, Yang XT. Artificial neural networks and thermal image for temperature prediction in apples. *Food Bioprocess Technol.* 2016;9(7):1089–1099.

45. Lu S, Wang X. Toward an intelligent solution for perishable food cold chain management. *Proc IEEE Int Conf Softw Eng Serv Sci ICSESS.* 2016;0:852–856.

46. Do Nascimento Nunes MC, Nicometo M, Emond JP, Melis RB, Uysal I. Improvement in fresh fruit and vegetable logistics quality: Berry logistics field studies. *Philos Trans R Soc A Math Phys Eng Sci.* 2014;372(2017):2014.

47. Mohebi E, Marquez L. Application of Machine Learning and RFID in the Stability Optimization of Perishable Foods. Third International Workshop on Food Supply Chains. 2014:1–18.

48. Badia-Melis R, Mc Carthy U, Uysal I. Data estimation methods for predicting temperatures of fruit in refrigerated containers. *Biosyst Eng* [Internet]. 2016;151:261–272. Available from: http://dx.doi.org/10.1016/j.biosystemseng. 2016.09.009

49. Draganić A, Orović I, Stanković S, Zhang X, Wang X. Compressive sensing approach in the table grape cold chain logistics. *2017 6th Mediterr Conf Embed Comput MECO 2017 – Incl ECYPS 2017*, Proc. 2017;(June): 0–3.

50. Zhao W, Dai W, Zhou S. Outlier detection in cold-chain logistics temperature monitoring. *Elektron ir Elektrotechnika.* 2013;19(3):65–68.

Chapter 6

Leveraging the potential of artificial intelligence in healthcare supply chain management

Sapna Yadav

Delhi Pharmaceutical Sciences & Research University, New Delhi, India

6.1 INTRODUCTION

Artificial Intelligence (AI) has recently emerged as an important tool to manage the complexities of the common business challenges. AI, and related technologies such as machine learning, robotics and stochastic optimization, is increasingly prevalent in business and life and are gradually being used in the field of healthcare. This technology is used for many therapeutic and research applications, including drug development, chronic illness management, medical services, and diagnostics. In terms of patient care, logistics for the healthcare system, and illness diagnosis, AI technology outperforms humans. Additionally, AI systems improve people's quality of life. For instance, AI today has achieved significant advancements in the detection and treatment of cancer, particularly breast cancer. Therefore, to boost cost savings, raise patient happiness, and satisfy staffing demands, hospitals, NGOs, and humanitarian groups in catastrophe scenarios are searching for AI solutions.(1)

In addition to currently used methods, including quantitative predictive models (such as fuzzy inference systems, regression, machine learning, and neural networks), computer-assisted methods, meta-heuristic algorithms (such as genetic, ant colony optimization algorithm, particle swarm optimization algorithm, etc.), simulation methods (such as discrete simulation and agent-based methods), and optimization-based decision support systems methods (mixed integer linear programmes, stochastic optimization, etc.), there are also new methods such as genetic algorithms.(2) As a result, this SI looks for fresh AI-based solutions to the difficulties raised.

Another industry with enormous potential for AI is healthcare supply chain management. This was particularly the case during the COVID era, when the healthcare sector faced unheard-of service pressures.(3) However, this made it easier to identify many supply chain gaps and the potential of AI to close them. It has been noticed that the healthcare supply chain significantly relied on manual procedures and person knowledge, lacking the resources or technology to identify methods to simplify, reduce costs, or gain

DOI: 10.1201/9781003462163-6

insights, as supply chain demands became a primary concern. Leaders are anticipating the effects on the future of the healthcare supply chain as the pandemic begins to stabilise. One thing has become clear: To boost supply chain automation and optimization, supply chain AI is required.

6.2 THE NEED AND ROLE OF ARTIFICIAL INTELLIGENCE (AI) IN THE HEALTHCARE SUPPLY CHAIN (HSC)

With the COVID-19 pandemic disrupting healthcare supply chains and leading to a lack of medications, vaccinations, and a rise in patient demand, it appears that AI techniques are required to address the issues in the networks' healthcare supply chain networks. The application of AI is becoming commonplace in many medical professions and disciplines. Healthcare stakeholders and medical professionals may identify healthcare problems and solutions more quickly and accurately thanks to AI, machine learning, natural language processing, and deep learning.(4) AI helps a variety of healthcare stakeholders, such as the clinical trial teams of physicians, researchers, or data managers. It has sped up the medical coding, search, and confirmation procedures, which are essential for starting and finishing clinical research. Among the many uses for AI are the analysis of medical imaging, cost reduction in pharmaceutical research, analysis of unstructured data, building of platforms for drug discovery, etc. AI in healthcare is expanding and Business Insider claims that this will lead to a broad variety of future professions in the country's healthcare industry.

6.3 COMMON AI TECHNIQUES IN THE HEALTHCARE SUPPLY CHAIN

"AI techniques" means algorithms, architectures, data or knowledge formalisms, and methodological techniques, which can be described in a precise, clean manner.

6.3.1 Medical supply spend analysis

Surgical, medical and drug spend add up to nearly 15% of operating expenses for a health system. Significant cost savings exist – if only health systems had the manpower to sort through the web of vendor matrices, tier-based pricing, rebates, etc. to identify those saving opportunities. Furthermore, pricing is often made intentionally vague by vendors, making it nearly impossible for a human to optimize. Yet studies show that there is up to a 17% spend reduction available by streamlining procurement practices, which can be undertaken most efficiently by artificial intelligence. An AI solution can use purchase order history from an Enterprise Resource Planning (ERP) to compare products based on cost and clinical outcomes and can then recommend optimal products based on cost and clinical quality. Plus, it can review vendor contracts for any limitations, and once product changes are approved, AI can update the ERP and any other punchout catalogs.(5)

6.3.2 Automated inventory management

Inventory management is a highly complicated process. Assessing stock levels, addressing backorders, managing recalls or expired products; it's hard enough for our human employees to simply acquire the necessary products to the patient and physicians, let alone optimize. AI algorithms can improve demand forecasts, optimize inventory levels based on purchase and usage history, and automate the management of expired and recalled products. Hospitals can reduce write-offs and save money by right-sizing their inventory and improving purchasing.(5)

6.3.3 Preference card standardization

Health systems have thousands of physician preference cards outlining supply expectations for a specific physician and a given procedure. In these cards are hidden opportunities for standardization and scale. But it is too time-consuming for a human to go through. AI can parse through all these cards, analyzing product mixes, to find standardization opportunities. After human review, it can even go back and update individual cards. Plus, AI can find all the duplicate or out-of-date cards to clean up and streamline the records.(5)

6.3.4 Three-way matching

On the financial side of the supply chain, AI can automate the three-way matching process. Currently, three-way matching is extensively manual, making it time-consuming and error-prone. AI can automate the entire process, reconciling invoices, purchase orders and receiving reports, to ensure that payments are accurate. It helps hospitals optimize their cash management, reduce supply payment errors, and save valuable employee time.(5)

6.3.5 Prescriptive analytics

Healthcare executives need a mechanism to get into the vast volumes of clinical and supply chain data in order to uncover trends that will enhance patient care and hospital finances. To aggregate data from all these sources, they require predictive analytics driven by AI. Insights and possibilities hidden in the webs of data will start to emerge as health systems continue to engage in AI across the company, eventually enhancing patient care and buying choices. For instance, in addition to Olive's other supply chain use cases, we are developing a process that can evaluate a hospital's present patient load in order to forecast supply demands.(5)

Predictive analysis, which is already well established in businesses of all sizes, has enormous promise for healthcare supply chain management. It can assist with a wide range of issues:

- Identifying which products and services improve outcomes.
- Contributing to the creation of enhanced inpatient care plans
- Examining the hospital's present patient load to project supply needs.

6.3.6 Machine learning models used in vaccine supply

There is a lot of discussion about how AI, like many other new technologies, might improve clinical, administrative, and financial performance as well as patient and physician experiences. At the same time, nothing has been said about what has to happen to make sure that artificial intelligence lives up to its potential.

One of the most ignored problems is data governance. AI has the benefit of analysing huge amounts of data to find patterns and underlying links that humans would require much more time to grasp.

Additionally, it enables users to supply several factors, such as those thought to have some bearing on the current scenario, to the artificial intelligence system. An intriguing model structure, called a Mixed Density Network (MDN), has been included into the basic framework of neural networks to address supervised learning issues where a single conventional probability distribution cannot accurately simulate the target variables. Instead of a straight prediction value, the network output in the stochastic MDN (Combination Density Network) model is a mixture of distribution parameters. The proposed MDN model combines distributions and Long-short-term memory (LSTM) layers. In this model, one or more distributions are given parameters via LSTM layers, which are then combined with weighting. Maintaining all pertinent historical information is essential since more verified instances might increase the risk of infection in upcoming populations.(6)

Compared to the stochastic LSTM/MDN and linear regression models, the deterministic LSTM model performed better. The real dataset's patterns, however, were better predicted by the stochastic model.

6.4 USAGE OF AI IN SUBFIELDS OF THE HEALTHCARE SUPPLY CHAIN

6.4.1 Task automation

AI of the robotic process automation (RPA) variety is being used more often in the healthcare industry, notably in the claims processing industry. RPA automates and standardises repetitive tasks using software robots, freeing up staff time for other, more valuable tasks. RPA is being used in the supply chain to automate contract-management tasks including pricing checks and adding contract requirements to sourcing strategies. Additionally, automation and AI are already being utilised to improve supply chain practises, such as proactive backorder management.(7)

- Tracking down medical alternatives.
- Automating updates to preference cards.
- Controlling invoices and point of sale (PoS).

6.4.2 Aids in cost-cutting

Evidently, the potential for long-term cost-effectiveness is one of the more alluring benefits of using AI in healthcare supply chain management. To enable hospitals and other healthcare institutions to employ machine learning to standardise supplies using treatment result data, enterprise solution integration is essential. Standardizing commodity supply using artificial intelligence also makes it possible to analyse massive volumes of data more thoroughly and compare the price and effectiveness of various products.(8)

Additionally, Healthcare Supply Chain Management can use AI to create and update standards in almost real time, enabling doctors to use techniques like outpatient treatment or skipping expensive tests if they haven't been shown to improve outcomes—especially when other factors have already brought costs close to the norm.(9)

6.4.3 Matching patient care demands

As more information about how goods perform in typical clinical situations becomes accessible, there is an increasing need to match the right product to the right patient. The needs of certain patient groups must also be considered while restructuring treatment strategies.(7)

By using this data for value analysis and procurement, AI may be crucial in establishing what is best for patient types and ensuring that the proper items are accessible where they are needed.

6.4.4 Substitutes during labor shortage

Along with the COVID epidemic, healthcare supply chain management also saw a growing nursing staff shortage. With the help of automation and AI, hospitals are able to increase capacity and do more with the staff they already have.

Employees are freed up to concentrate on more strategic, high-level tasks or patient care thanks to artificial intelligence. Additionally, artificial intelligence may work in tandem with human workers to improve the efficiency and accessibility of their duties.

Together, AI and machine learning help the workforce operate more quickly and intelligently while also boosting morale and reducing burnout.(10)

6.4.5 Supplier management data

Contrary to retail and other businesses, disruptions in the healthcare supply chain management may be quite worrying. Artificial intelligence may help manufacturers gather data, assess inventory status, and help suppliers anticipate backorders. They can also aid in navigating the intricate supply chains so that delays are better anticipated, remedial action is taken, and assistance is given to customers in finding alternatives.(11)

6.4.6 Optimizing logistics support

In order to plan the shortest ambulance routes for transferring patients to hospitals or other care delivery facilities, AI-enabled patient flow companies can make use of methods frequently used by third-party logistics businesses. The healthcare supply chain staff may also employ these similar technologies to assist with the transfer of care outside of critical care facilities.

The most effective modes, frequencies, and routes for moving supplies and caretakers to the continually increasing number of locations where they will be needed may be determined with the help of artificial intelligence. These locations might include everything from residences and clinics to ambulatory surgery facilities and emergency rooms.(12)

6.4.7 Improved data governance

There is a lot of discussion about how AI, like many other new technologies, might improve clinical, administrative, and financial performance as well as patient and physician experiences. At the same time, nothing has been said about what has to happen to make sure that artificial intelligence lives up to its potential.(13)

One of the most ignored problems is data governance. Artificial intelligence has the benefit of analysing huge amounts of data to find patterns and underlying links that humans would require much more time to grasp.

Additionally, it enables users to supply several factors, such as those thought to have some bearing on the current scenario, to the artificial intelligence system.

6.5 BARRIERS TO THE ADOPTION OF AI IN THE HEALTHCARE SUPPLY CHAIN

The infrastructure, early investment, and high cost of using AI in healthcare are all major roadblocks in delivering such systems. The infrastructure needed for AI to develop in India is still insufficient. There are several obstacles to utilising big data, which is essential for AI-driven healthcare, including: the sheer volume of unstructured data sets and interoperability issues; the lack of open medical data sets; insufficient analytics tools capable of handling big data; and worries that algorithms may produce data that reflects cultural biases.(14)

Acceptance and the use of AI are sometimes hampered by trust concerns and fear about new technology, especially among the elderly. Both medical professionals and the general public still do not fully comprehend AI's advantages. Adopting AI in healthcare is significantly hampered by the poor framework for ensuring the privacy, security, quality, and accuracy of AI solutions.(15)

Uncertainties and regulatory shortcomings continue to be problematic. A significant barrier for start-ups in AI is the requirement for proof of acceptable outcomes in the form of expensive and time-consuming clinical studies. Another important problem that must be addressed is responsibility for AI, which now only applies to the clinician and not the technology. Lack of trust might be exacerbated by worries about human job losses.(14)

Lack of trust might be exacerbated by worries about human job losses. However, AI is frequently believed to close the supply-demand gap and serve as a doctor's helper in the healthcare industry. A significant obstacle to implementing AI in healthcare is the dearth of skilled professionals in the field.(15)

The underrepresentation of minority groups in the data used to develop algorithms and solutions, the predominance of men in the software industry, which results in a bias against men in technology, and the greater benefits to higher income populations with access to technology are all inequality concerns in the adoption of AI in healthcare.

6.6 THE IMPACT OF COVID-19 ON PUBLIC HEALTHCARE SUPPLY NETWORKS

The COVID-19 pandemic highlighted the importance of supply networks for healthcare, exposing their flaws and gaps. Without spending astronomical expenses, the sector may employ several techniques to reduce supply chain interruptions during big events. Every country, company, and supply chain on our earth have been touched by the remarkable COVID-19 pandemic. With hospitals on the verge of closing due to capacity overflow, vital item supply chains disrupted, and federal and state authorities failing to take palliative and preventative measures, the pandemic threw the healthcare system into crisis. While there were crisis plans and stockpiles in place for governments and private sector enterprises, the pandemic uncovered numerous significant supply chain weaknesses, including a lack of personal protective equipment (PPE) and testing kits.(3)

Of course, a variety of things might cause supply chain interruptions, including natural catastrophes, war or terrorist attacks, supplier bankruptcy, labour conflicts, cyberattacks, and data breaches. The COVID-19 pandemic differs from previous ones in terms of the degree of unpredictability, the duration of the disruption, and the simultaneous impact on several geographical locations. Additionally, unlike many earlier disruptions, COVID-19 has had an impact on both the supply and demand for goods and services.(3)

Organizations from a wide range of industrial sectors have tried to stabilise their supply chains in reaction to the pandemic by performing risk analyses and putting in place business continuity plans. To adapt to shifting consumer expectations, many have expanded their product lines and created new items using their current resources. For instance, some producers of textiles started producing PPE, and some distillers started producing hand sanitizer. Other companies have improved their supply chains by using 3D printing technology to produce goods more quickly in response to customer demand. Finally, several have underlined the necessity of nearshoring or bringing industrial facilities back onshore.(3)

It is crucial to first recognise the difficulties that led to the significant supply chain disruptions observed during the pandemic in order to improve and stabilise the healthcare supply chain in future. The next step is for pharmaceutical and healthcare enterprises to evaluate which solutions will assist them prevent supply chain interruptions during severe catastrophes without incurring astronomical expenditures. For instance, while maintaining large quantities of safety stocks for a range of healthcare products and/or reshoring manufacturing of several goods might increase resilience, both solutions would be prohibitively expensive and thus impractical. Finally, the private sector cannot provide all of the remedies. Public health requires emergency preparedness, and federal, state, and local governments must evaluate what policy recommendations they should make in the aftermath of this event.

The sluggish reaction of the healthcare supply chains to the COVID-19 disaster was caused by several issues. These include export bans implemented by nations where protective clothing, medical equipment, and pharmaceuticals were produced, which limited supply to importing countries; panic buying and the stockpiling of critical supplies, which led to shortfalls in supply; and not having a sufficient number of workers to produce and transport products due to workers being absent or failing to show up for work. A lack of transparency in evaluating the current centralised supply, such as the Strategic National Stockpile (SNS) maintained by the U.S. Department of Health and Human Services, and Poor alignment and coordination among federal, state, and local agencies, as well as healthcare organisations, resulting in a fragmented approach to ordering and fulfilling orders are some of the problems.(16)

These flaws influenced clinical treatment, leading to limited testing capacity, poor care coordination, and supply restriction. Important parties have picked up on this issue. Even the World Health Organization issued a statement on March 5, 2020, cautioning that interruptions in the PPE supply chain globally left medical personnel gravely underprepared to handle the pandemic.

The fact that the healthcare system has long tried to contain supplier costs in order to balance out declining payments from Medicare and Medicaid as well as private insurance plans is one of the main reasons for the system's difficulties. Many medical manufacturers have been moved abroad to reduce costs so they may benefit from free labour and tax breaks. Over time, this has made the United States unduly reliant on foreign production for many basic healthcare products. For instance, China is the country where 80% of all face masks are produced. As the pandemic expanded, this reliance had significant effects. When the majority-producing nations for face shields and masks halted production and imposed export prohibitions, the U.S. struggled to get enough of these goods. Hospitals also find it challenging to obtain precise information about current production levels or incoming shipments due to the worldwide nature of healthcare supply chains, which limits their capacity to predict future shortages.(16)

6.7 CONCLUSION

The healthcare sector has seen several procedures thanks to digitalization. It has the potential to enhance logistics through analytics, forecasting, stock taking, inventory backorders, data governance, and labour substitution with its debut in the field of healthcare supply chain management. Any company or organisation that wants to integrate artificial intelligence into their healthcare supply chain management system must carefully consider their goals and put solutions in place.

6.8 FUTURE SCOPE

Governments, corporations, and investors all around the globe have prioritised healthcare thanks to COVID-19, which has also driven attempts to use artificial intelligence to improve our health, from medication research to more effective hospital operations to better diagnostics.

The future of AI in healthcare: An introduction of machine learning, artificial intelligence (AI), and natural language processing (NLP) with a focus on the healthcare sector. Impact on patients, doctors, and the pharmaceutical business of current and future healthcare applications. An examination of how AI in healthcare could develop as new technologies change how medicine and healthcare are provided over the coming ten years.(17)

REFERENCES

1. A set of efficient heuristics and meta-heuristics to solve a multi-objective pharmaceutical supply chain network | Semantic Scholar [Internet]. [cited 2023 Jul 13]. Available from: https://www.semanticscholar.org/paper/A-set-of-efficient-heuristics-and-meta-heuristics-a-Goodarzian-Kumar/788f0a0d799dab1275e2bd612b4ce13401a0a018

2. Wisetsri W, Vijai C. Rise of Artificial Intelligence in Healthcare Startups in India. 2021 Mar 1.

3. COVID-19: Impact on Health Supply Chain and Lessons to Be Learnt - Anu Sharma, Pankaj Gupta, Rishabh Jha, 2020 [Internet]. [cited 2023 Jul 13]. Available from: https://journals.sagepub.com/doi/full/10.1177/0972063420935653

4. The rise of artificial intelligence in healthcare applications - ScienceDirect [Internet]. [cited 2023 Jul 13]. Available from: https://www.sciencedirect.com/science/article/pii/B9780128184387000022

5. 5 ways AI and automation are optimizing healthcare supply chains in 2021 [Internet]. [cited 2023 Jul 15]. Available from: https://oliveai.com/resources/blog/5-ways-ai-and-automation-are-optimizing-healthcare-supply-chains-in-2021

6. Future of artificial intelligence and its influence on supply chain risk management – A systematic review - ScienceDirect [Internet]. [cited 2023 Jul 15]. Available from: https://www.sciencedirect.com/science/article/abs/pii/S0360835222002765

7. www.ETHealthworld.com | Pharma. [cited 2023 Jul 15]. How AI is helping the supply chains in the healthcare sector - ET HealthWorld | Pharma. Available from: https://health.economictimes.indiatimes.com/news/health-it/how-ai-is-helping-the-supply-chains-in-the-healthcare-sector/91232124

8. The Role of Automation and AI in Healthcare Supply Chain Management [Internet]. Aeologic Blog. 2023 [cited 2023 Jul 15]. Available from: https://www.aeologic.com/blog/the-role-of-automation-and-ai-in-healthcare-supply-chain-management/

9. How AI in healthcare supply chain management (SCM) can cut costs [Internet]. 2019 [cited 2023 Jul 15]. Available from: https://spendmatters.com/2019/09/09/how-ai-in-healthcare-supply-chain-management-scm-can-cut-costs/

10. www.ETHealthworld.com. [cited 2023 Jul 15]. Transforming Hospital Supply Chain with AI and Automation - ET HealthWorld. Available from: https://health.economictimes.indiatimes.com/news/industry/transforming-hospital-supply-chain-with-ai-and-automation/87976757

11. Kumar A, Mani V, Jain V, Gupta H, Venkatesh VG. Managing healthcare supply chain through artificial intelligence (AI): A study of critical success factors. *Comput Ind Eng*. 2023 Jan 1;175:108815.

12. Artificial intelligence in supply chain management: A systematic literature review - ScienceDirect [Internet]. [cited 2023 Jul 15]. Available from: https://www.sciencedirect.com/science/article/pii/S014829632030583X

13. Deloitte Insights [Internet]. [cited 2023 Jul 15]. Smart use of artificial intelligence in health care. Available from: https://www2.deloitte.com/us/en/insights/industry/health-care/artificial-intelligence-in-health-care.html

14. Haider H. Barriers to the adoption of Artificial Intelligence in healthcare in India.

15. Barriers to Artificial Intelligence Adoption in Healthcare Management: A Systematic Review by Mir Mohammed Assadullah :: SSRN [Internet]. [cited 2023 Jul 13]. Available from: https://papers.ssrn.com/sol3/papers.cfm?abstract_id=3530598

16. A sustainable-resilience healthcare network for handling COVID-19 pandemic - PubMed [Internet]. [cited 2023 Jul 13]. Available from: https://pubmed.ncbi.nlm.nih.gov/34642527/

17. Artificial intelligence in the healthcare supply chain | Merck [Internet]. [cited 2023 Jul 15]. Available from: https://www.merckgroup.com/en/research/science-space/envisioning-tomorrow/precision-medicine/ai-in-supply-chain.html

Chapter 7

Artificial intelligence-assisted telemedicine

A boon in healthcare

Sheetal Yadav, Jaseela Majeed and Sheetal Kalra

Delhi Pharmaceutical Sciences & Research University, New Delhi, India

7.1 INTRODUCTION

The term telemedicine has been coined to refer to the delivery of clinical services to patients through the application of information communication and technology. The provision of telemedicine functions is achieved by catering to the patient's needs through a combination of calls and video conferencing. Globally, physicians use telemedicine for digital imaging transmission, video consultations, and remote medical signals (Wootton et al., 2017). The origin of telemedicine has been as a response to the lack of the time required for effective consultations. The term "telemedicine" is frequently used in conjunction with "virtual consultation", which is the term given to the use of electronic information and telecommunication technologies to provide support and promote long-distance clinical healthcare, for the medical education of clinical professionals and to assist in the maintenance of public health and its administration (Lilly et al., 2017). A distinction can be drawn between the two terms, however, since telemedicine can be regarded as the purely clinical application of the information technology in healthcare whereas virtual consultation is used as an umbrella term for nonspecific healthcare-related activities. With the advances in technology and their application in a diverse range of fields, evolved concepts such as telemedicine have been successful in their global implementation. The application of telemedicine has been considered to be useful in both developed and developing nations. Core benefits of the application of integrated telemedicine in the healthcare system are inclusive of increments in the revenue, maximization of patient reach, provision of convenience, increased feasibility, high diversity and availability of several healthcare professionals on a single platform and with improved healthcare quality (Mehrotra et al., 2016).

7.2 EVOLUTION AND FUNCTIONING OF TELEMEDICINE SERVICES

Mobile health and telemedicine services make use of telecommunication services and have therefore been directly linked with technological evolution. The initiation of telemedicine can be associated with the 1950s Nebraska Psychiatry Program (Abrams, 2014), which had made use of two-way closed circuit televisions (CCTVs). Multiple telemedicine projects have been initiated in America since this date, for consultation and neurological assessment along with speech therapy and also the training of staff members. Teleconferencing equipment has also been used in telemedicine. Healthcare projects that used telemedicine are also associated with the usage of mobile medical examination rooms and satellite communication (Agboola et al., 2014). The widespread use of computers and intelligent devices also enhanced the use of telemedicine in the developed countries in the 1980s. Telemedicine and virtual consultation have also been described as electronic health and eHealth in certain studies. The elements of eHealth include electronic patient records, business transactions, staff training, and health insurance. However, there has recently been a major hike in telemedicine health services. The early mHealth applications were developed in combination with notebook computers and connected via the internet. Cellular communication and mobile computing had operated with multiple networking protocols (Ciere et al., 2012). The initial telemedicine systems recorded electrocardiograms, through the use of pocket computers. Through the use of mHealth applications and network systems, the use of telemedicine has further gained in popularity. The initial mobile health and telemedicine applications focused on oxygen saturation levels, pulse oximetry, ECG, blood pressure, movement, temperature sensors, and continuous monitoring (Amit & Zott, 2012). The use of wearable sensors, electrodes, and connection fabrics has also evolved in order to be used in telemedicine with advances in technology (Darkins et al., 2014). This is in contrast to the traditional medical devices which were conventionally used in the market. The US Food and Drugs Administration (FDA) had classified the telemedicine and mobile health services into different categories. These categories included remote health monitoring services, disease management, and the interface/extension of existing services, and telemedicine had also assisted in the diagnosis, sports and physical fitness, healthy living, prompts and alerts, performance-enhancing apps for healthcare providers, and reference apps (Abrams, 2014). The remote health monitoring apps assisted in the monitoring and assessment of the body's vital readings such as heart rate and ECG for the patients. The disease management apps provide telemedicine services that included the measurement and logging of blood pressure, blood glucose levels, expiratory volume, etc. The extension and interface services provide blood pressure monitoring services along with temperature measurement. The telemedicine and mobile health services that are categorized as assistance in diagnosis focus on tele-radiology, tele-cardiology, and the review and assessment of the laboratory test results. The sports and fitness applications measure the physical activity and vital signs, such as the users' heart rate. The mobile health apps categorized as healthy living apps can include both diet plans and exercise logs. Various virtual consultation applications also focus on the setting of reminders and alerts. These apps assist in appointment generation, maintenance, and medication reminders. Telemedicine and eHealth applications have also diversified to function as productivity enhancers for the healthcare providers. These applications aim to do so by assisting in the scheduling and assessment of the patients' health records. Medical guides have also been developed by the telemedicine and virtual consultation service providers which are included in the category of reference applications for healthcare by the FDA (American Telemedicine Association, 2014).

The application of telemedicine has been divided according to their utility and application. These include phone-based consultations, interactive video conferencing, store and forward technology, and remote monitoring. The phone-only consultations have been one of the oldest methods of virtual consultation and are still prevalent today in the large parts of the world where there is limited access to healthcare services. Local and indigenous telephonic networks, such as integrated services for the digital network (ISDN), are used for the transmission of telephonic calls, photos, and videos simultaneously for virtual consultation consultations and services (Burke & Hall, 2015).

Telephonic consultations are more common than video consultations and have been used regularly for patient–clinician communication. Two-way communication is possible in the virtual consultation services and telemedicine sector through the use of video conferencing. Video conferencing is used for the real-time two-way transmission of the digitized pictures for virtual consultation and telemedicine. Store and forward technology is also used frequently for telemedicine (Doarn et al., 2014). Digital images are taken at the site of the patient for care and are forwarded to the clinician at another site for interpretation. This form of telemedicine is often used in radiology and dermatology. The process is often asynchronous and is used in certain consultations. Remote monitoring is also often used in the telemedicine technology. Remote monitoring is used when the technological devices are used to remotely gather the data of the patient by monitoring and send it to a monitoring station for interpretation and analysis. This is inclusive of vital sign assessment, passive medical observations, the use of alarm systems, and the provision of care support for self-management (Desko & Nazario, 2014).

7.3 THE ROLE OF AI AND DDT IN HEALTHCARE

Emergency mobile communications and network technologies have also emerged to provide extensive support for healthcare support for the patients. The wireless telemedicine systems and services have also gained popularity within the emergency health departments and care departments. These have also been useful for assistance in hospital and care services in remote locations. In cases of severe head injuries, serious spinal cord injury, and injuries to internal organs, the way of transportation and the way of provision of care play crucial roles in the recovery of the patient. Most of the telemedicine and virtual consultation applications that are used directly by the user operate by using the wireless biosensors; these form a "Body area network" that reports the user's vital signs (Bozikov, 2015). In emergency healthcare procedures, the biosensors are used for sensing the surgeon hand movement in real time with high precision to prevent clinical errors. The cameras are commonly used in telemedicine to transfer patient images without any obstruction during a medical procedure and surgeries to provide the best care by effective monitoring. The actuators are also often used in the clinical settings to provide effective movement. The effectiveness and mode of operation of the telemedicine service have been extremely dependent on technological advances. Strong communication systems and improvements in them have allowed for increased reliability in the healthcare procedures by making them more secure, technologically sound, and less error-prone. Telemedicine application through strong communication networking has also allowed for information sharing between the healthcare practitioners and the care providers, thereby improving access and feasibility. Effective management of the databases with more organized information on patient records and care need has also been established by the effective use of telemedicine. This data is highly important since it can help in the prediction of trends, the analysis of health records, and also in the research and development of healthcare. With major advances in the field and high applicability, the telemedicine market is growing rapidly with a huge number of global consumers. Both the popularity and the awareness of telemedicine grew when NASA used telemedicine and virtual consultation services to monitor the health of astronauts in the 1980s (Burke & Hall, 2015).

Telemedicine services have been an effective way to control the costs of healthcare. Telemedicine saves money by limiting the number of in-person doctor visits, immediate ER costs and hospitalizations for minor illnesses and also through the management of patients' chronic illnesses. Therefore, it serves as an effective measure of cost reduction in healthcare systems. The telemedicine service has been recognized for effective cost saving. For instance, the Veteran Affairs (VA) telemedicine program was conducted in the United States, where the participants recorded significantly lower health costs in comparison to the those who were not enrolled in the telemedicine program (Whittaker & Wade, 2014). It has also been observed that telemedicine and virtual consultation problems are extremely effective in reducing the care costs of the patients with chronic health conditions that require management. It has also been found to be effective in reducing the mortality rates of patients suffering from chronic illnesses. It has been evidenced, for example, through the study of Chronic Obstructive Pulmonary Disease

(COPD) patients that introducing virtual consultation services saved an average of $2,931 a year because of the decreased usage of office visits, fewer hospitalizations, and also shorter hospital stays (De San Miguel et al., 2013). This reduction in cost was associated with limited ER visits and hospitalizations for the people who had participated in the telemedicine program. The effectiveness of telemedicine in cost reduction is also associated with minimizing the overall healthcare burden and reducing the indirect costs of healthcare services on the consumers. Further, since telemedicine deploys advanced technology for diagnosis, management, and care facilitation, the health costs, in the long run, are also reduced by saving time. Use of the advanced technology also minimizes the operational costs of care. The cost of availing the telemedicine services is less than the average cost of visiting specialists in healthcare settings. It has also been recorded that patients often record a high degree of satisfaction with their medical services, thereby reducing the frequency of visits to the hospital and helping in the reduction of care costs (Cafazzo et al., 2012). Since telemedicine helps in the effective management of chronic health conditions, it also contributes to the decrease of the rate of hospitalizations and readmissions which also contributes towards the reduction of care costs. Telemedicine services also helps in assisting patients to adhere to their medication by providing constant reminders and assessment of the measurable outcomes where they can witness their health improvement (Chen et al., 2013). Therefore, the application of telemedicine helps in the management of health conditions; this prevents the cost that may arise due to secondary complications because of medication non-adherence and, therefore, helps in minimizing the health costs that are caused by the increased burden of medication non-adherence. Additional costs that are saved through the application of telemedicine include reduced travel time and there being no need to take time off work in order to access healthcare services. This is of significant help for rural populations where individuals might be unable to travel to healthcare facilities given financial or time constraints (Hamel et al., 2014).

The epicentre of telemedicine revolves around its ability to increase the access and ability of medical care to reach the remote and rural population (Holeman et al., 2014). Telemedicine services have also been shown to be effective for people who use the medical services infrequently and especially to the communities or individuals who may regarding accessing healthcare services as somewhat stigmatized or who were restricted through the difficulty in accessing outreach services (Howitt et al., 2012). The telemedicine is often used in complementation along with the primary care models that use an integrated approach to deliver an overall better level of care and healthcare facilities to people living in a diverse range of communities and with variable access and affordability to the healthcare services. Since telemedicine helps in the organization and also in the record keeping of the medical services, it also helps in the identification of trends and development of institutionalized practices with a targeted approach to improving the overall care access among the people (Johnston et al., 2015). The use of telemedicine also helps in updating the care guidelines of medical facilities as a large number of people can be targeted, their needs can be effectively understood and denoted, and the obligatory amendments to make the virtual consultation-care and telemedicine services more effective can be performed. The telemedicine also assists in the improvement of healthcare by directly providing care services to patients, enhancing communication between the patient and the healthcare service provider and also helping in the development of a "patient-centred approach" where each patient is dealt with an individualistic manner for the complete care (Klonoff, 2013). Since the use of telemedicine helps in the development of flexibility between the clinical care services provided from the patient perspective and clinician perspective it also allows for more effective care (Baker et al., 2013).

Those healthcare professionals who use telemedicine in healthcare are also able to achieve higher precision in the provision of care services. Through the digital transfer of data, the patient's needs could be addressed by specialists and multiple recommendations may be generated for complex cases that may require assistance and consultation (Henderson et al., 2013). Further, in surgical procedures the use of sensors and camera graphics have also significantly assisted the medical professionals to undertake guided procedures to reduce the chances of error and improve the general quality of healthcare. Telemedicine has also helped in the improvement of quality of care by providing the access to healthcare in the remote locations where the medical facilities were unavailable. The healthcare structure of a state or of a country is dependent on both rural and urban healthcare systems and therefore, to improve the overall healthcare, it is crucial to improve the reach and quality of the healthcare services in rural and remote locations (Wootten, 2012).

The use of telemedicine in the improvement of quality of the healthcare facilities has been foundational. Use of modern diagnostics and the virtual consultation services have allowed for an improvement in overall health infrastructure. Since it is possible for the healthcare practitioners to have accessibility to the patient and disease information via the organized databases, it has become easier for healthcare professionals and health researchers to identify disease trends and outbreaks, and to regulate them with improved efficiency. This has also assisted in the development of appropriate healthcare programs that have been personalized for people and societies to ensure the highest level of quality healthcare (Baker et al., 2013). Since the patient's needs are effectively recorded and communicated through telemedicine application, the improvement in the quality of healthcare has been observed at multiple levels in the healthcare structure. The diagnostic services have improved with the application of telemedicine since the use of advanced technology can help in the prediction and suitable identification for the healthcare facilitator. Diagnostics have also improved as the primary care provider can also take assistance from the specialists for consultations.

The care provision has been improved by direct interactions with the patients in a highly personalized approach which ensures patient-centred care and an effective addressing of patient needs. Since both the time required and the costs of the healthcare services are reduced by the application of telemedicine, more resources may be spent on effective planning, research and development and quality improvement in the healthcare services for patients (Bashshur et al., 2014).

Strengths	Weaknesses
• Ease of access to healthcare services for remote patients • Saves traveling time • Saves patient's money • Reliability of telemedicine is good • Prevents worsening of patient's condition • Provides desirable results • Readily and willingly acceptable to patients • Reduces patient traffic in tertiary care hospitals • Feasible and convenient to provide • All the applications of telemedicine are beneficial to patients	• Cannot replace face-to-face consultation • High cost of equipment, maintenance and technology • Lack of necessary training/skills for use of equipment • Inadequate infrastructure with adequate bandwidth • Lack of user-friendly softwares • No well-identified business model to ensure sustainability • No benchmarking and evaluation schemes • No best practices • Lack of standards for patient safety
Opportunities	Threats
• Advancement in ICT • Can overcome healthcare manpower shortages • Limited competition – gigantic opportunity to invest • Can be implemented in all hospitals equipped with internet facility • Good compatibility between IT professionals and clinicians • No fear to lose jobs if telemedicine is deployed	• Resistance to change • High power cuts in rural areas • Illiteracy • Poor internet connection in rural areas • Social and cultural barriers towards ICT utilization • Concerns regarding patient privacy/ confidentiality

7.4 CHALLENGES TO THE IMPLEMENTATION OF AI IN THE INDIAN HEALTHCARE SYSTEM

The rise in the availability of tele-health services has been regarded as a boon to the accessibility of expert specialty care services across the world. However, the availability of telemedicine in India has been largely concentrated and restricted. Some major gaps are associated with the application of telemedicine in India. There is a dearth of evidence-based research to substantiate the success and reach of tele-health care across different districts of India (Meher et al., 2017). Further, there is limited information available about the perceptiveness of the concept, its awareness, and its utility. It is crucial to possess information about the acceptance of technology in demography to predict its success. However, negligible research has been done in the direction of telemedicine to affirm the utility, application, and success of the technology in the region. Telemedicine in India is provided by several companies, including NetMeds, WelcomeCure, Practo, MedGenome and several others. However, a market analysis of their reach, the population demographics of the users and the availability of these services continue to be a blind spot in the research (Ahmed et al., 2019). There is also a significant gap in the evaluation of the management strategies used by these establishments for their operations. No relevant research exists to evidence the lack of propagation of technology even when it was introduced about two decades ago, making it an over-promising and underachieving goal for the population of India. These existing lacunae have limited the market research and analysis associated with telemedicine and therefore is of prime significance for prospects (Mehrotra et al., 2016).

Even when the primary asset of the tele-health application is its cost-effectiveness, the initial costs of building a telemedicine health structure can be cost-intensive. Therefore, the study is required to minimize the initial operational costs so that countries and states with limited resources can also apply the technology to improve the overall healthcare structure and system of the region (Gellis et al., 2012). Therefore, further analysis to measure the quality of medical care outcomes and influence on costs is required. It is also vital to measure the quality of telemedicine at different levels and find the lacuna to make the services even better. Although the current literature asserts that the implementation of telemedicine is effective in the upgrading of the overall quality of the healthcare structure, it is crucial to find out its bearing on the various parameters of healthcare which include database management, diagnostics, patient–clinician interactions, and health improvement status (Grabowski & O'Malley, 2014). Determination of these specific parameters will allow for the identification of the highest effective niche of telemedicine and also help in the improvement of the overall telemedicine structure. This will also be helpful in the implementation of a tele-health system in various areas and societies based on the priority needs (Labrique et al., 2013). The tele-health interventions can be globalized to have a broader reach in the societies and to have a broader access among the people who have inadequate access to telephones and internet availability. The use of telemedicine in contrast to in-person care is larger; it can be used to reach individuals with special needs or disabilities. Therefore, it is critical to explore the medicine market needs of the population with disability to underpin the tele-health needs of the community in order to grow the telemedical services across the globe (Mehrotra et al., 2013).

The study of tele-health and its application has been also linked with several paradoxes. The business model paradox associated with telemedicine asserts that, despite several predictions concerning the growth of telemedicine, it is not yet accepted as the standard approach to global health sector activity. Even when the use of telemedicine has been introduced, such as in the healthcare systems of the United States, the dominant setup remains face-to-face interviews (Minatodani & Berman, 2013). In-person patient care and hybrid models associated with the telemedicine application, along with traditional health services, have remained popular amid the commonalities. Even when the telemedicine and tele-health care services are largely supported, the growth of this sector has been limited, creating a considerable gap in the present utilization of the tele-health services with the vision of the health promotion established for the telemedicine application. For the successful application of telemedicine in the sphere, it is important to understand that there is a necessity to move the focus of healthcare systems from the prevention of disease to capacity building in the healthcare sectors. It is also vital to adapt the healthcare provision to the community involved by ensuring that health is perceived as an everyday resource (Pearl, 2014). It is also critical to comprehend that health promotion is not only restricted to specific approaches but also moves beyond the healthcare structure and its delivery system and penetrates into the social and cultural beliefs of the community or the individual involved. Therefore, empowerment and equality through telemedicine application in the healthcare sector should be promoted. Telemedicine can serve as a tool for the empowerment of the communities as it functions as a "process of enabling people to escalate control over the healthcare system and also to better their health" (Pratt et al., 2013). Health promotion through the application of telemedicine can also function by moving the health out of the expert action frame and making it popular in the community action through a participatory approach. Telemedicine can also function in the area of health promotion by enabling the communities, and advocating and mediating the need for good health. However, the current telemedicine models focus more on the disease management than on prevention and health promotion (Rowell et al., 2014). There is also limited substantiation regarding the resource modeling and transaction mechanisms of healthcare systems (Tamrat & Kachnowski, 2012). The value structure and organizational modeling of tele-health are also pervasive in its application. This gap is present in telemedicine as there is a dissimilarity in the application of telemedicine models with high inconsistency. The common domains of telemedicine businesses include transactions, marketing, and back-end operations (Smith, 2013). The transaction in telemedicine is delivered by expert and skilled diagnosis and consultations provided through telephonic conversations or video conferencing. The marketing in this sector functions through communications and virtual consultation interactions with the patients. The clinical data maintenance and administration are handled at the back-end operations in telemedicine services (Smith et al., 2012).

Another underlying problem with the application of telemedicine and a prevailing gap in the existing literature submits that most of the barriers in its application have remained constant. Healthcare systems are a complex amalgamation that requires complex product system dynamics with social product systems. It is important to develop trust in the medical practice through collaboration and knowledge sharing. Even when telemedicine promises to reach out to the remote locations and inaccessible locations related to the healthcare services, people also express distrust in the usage of technology-based services against in-person or face-to-face consultations in healthcare, keeping the major hindrance in the application of telemedicine intact (Mehrotra et al., 2013). The adoption of information technology in the healthcare sector has been uneven and is connected with the electronic record keeping and database management systems. Barriers in the embracing the telemedicine for medical care have also been associated with its application in the multidisciplinary team collaborations. The healthcare systems are spread based on the component subsystems that are independently structured resulting in the uneven adoption of the technology and collaboration with the virtual consultation services (Kayingo, 2012). It has also been argued that the many sectors of the telemedicine health services are disorganized in terms of the adoption of the innovation system integrations. Lack of clinical involvement in the adoption of telemedicine has also remained a major hindrance in its successful adaption in the healthcare systems. It has been considered that clinicians often look for conventional approaches as an alternative to face-to-face patient interactions than adopting the virtual consultation care services (McGowan et al., 2012). "Lack of information, lack of information handling skills, lack of time and alleged peripherality to the job" are the major barriers which are associated with the application of telemedicine by the rural healthcare practitioners; this impairs the application of the telemedicine and allied services (Law & Wason, 2014). Skepticism about the application of the technology has also been prevalent in large sections of the communities and countries directly related with the limited spread of the technology. The clinician's knowledge about technological advancement and communication technologies also influences the application of telemedicine. The user-friendliness and the approach of telemedicine applications is, thus, an important factor which determines its adoption in the mainstream healthcare system (Zickuhr & Smith, 2013). In conditions where the patients possess more than one clinical condition, it is required for the patient to use more than one application, making it a less preferable choice and affecting the sustainment of the telemedicine applications in the healthcare market. This also affects the user engagement with applications and creates a shortage for the tele-health applications user market. The use of language in the tele-health application also impacts the usage and effectiveness of the application. Therefore, there is a need for better strategic planning, the identification of the key targets, and project management of the tele-health services for the effective application of the technology for it to reach to a larger population and achieve its full potential (Law & Wason, 2014).

The lack of supportive frameworks and adequate policies for the application of telemedicine are also significant barriers to its effective application. The commercialization of telemedicine is hindered by the lack of availability of adequate resources; this serves as a market barrier for the telemedicine applications (Armfield et al., 2012). The application of these services is also impacted by the government policies and regulation as well as the payment systems established (Henderson et al., 2013). The cost of implementation and the adoption of telemedicine is inclusive of the technology and also the service costs and is established as a major obstacle in the implementation of the technology for implementation in the initial states. The government and the insurance companies do not ascertain the virtual consultation services for the reimbursements, making the users reluctant for its adoption and acceptance. Even when the cost saving is deliberated as a key asset of the telemedicine services, the preliminary costs of the implementation remain a major barrier that constrain its application and introduction in the healthcare systems globally (Wootten, 2012). Lack of effective research on the application and organization of tele-health serves as a major barrier as it fails to deliver the requisite rigor for its application and introduction in the new business markets hindering its expansion. Lack of available and adequate evidence regarding the effectiveness of the telemedicine and tele-healthcare facilities services also considerably impacts its adoption of the services and programs (Veit et al., 2014). Tele-health in research is often studied as a "pilot project" rather than as a long-run initiative impacting the health system of the community or geographical region under consideration (Wakefield et al., 2012). The effects of technology-based medical services are rarely studied in a holistic fashion. The implementation of healthcare services such as telemedicine have multiple modalities that require consideration. These include economical dimensions, technological dimensions, and social and ethical dimensions, that are often ignored in the majority of studies which largely focus on the clinical and economical frontiers of the tele-health application in the community (Istepanaian & Zhang, 2012). The lack of sufficient data to support the research and provide the substantiation to support the use of telemedicine in a continuous workflow for the effective supervision and treatment of patient conditions also severely impacts the application of the technology and the difficulty in building trust among the clinicians who use it (Whittaker & Wade, 2014). Lack of analysis on the logistics and the human resource constraints associated with the telemedicine applications is also directly associated with its application and determination of the implementation of the technology in the health sector. The researches available are also essentially devoid of patient experiences and emphasize the applicability and reach of tele-health services. The limitations of the telemedicine associated with the lack of end-user engagement has been addressed in most of the studies which have been published in regard to the implementation and experience associated with telemedicine (Wade et al., 2014). The literature regarding the telemedicine application has been dedicated to the perspectives of the clinical practitioners, virtual consultation experts, and health care professionals with limited reporting on the aspects of patient satisfaction and overall patient support towards the embracing of technology in clinical and non-clinical settings making the current research contradicting the "patient-centred" approach of telemedicine (Zickuhr & Smith, 2013).

7.5 FUTURE DRIVERS AND PROSPECTS OF TELEMEDICINE IN INDIA

India is a developing country with a massive population. The country's healthcare system is fragile and there is an impending shortage of healthcare workers in the country. Further, the accessibility and affordability of healthcare services to the rural population is also a major concern. In such circumstances, the introduction of telemedicine may serve as a boon to the country. The introduction and application of telemedicine across India would not only ensure the availability of better medical care services but also assure massive financial growth. India is regarded as one of the "young countries" of the world, where the majority of the population is of working age. This population has only limited time to spend on in-person consultations and could therefore serve as ideal customers for a telemedicine-based system (Pandya et al., 2019). Another positive prospect of telemedicine introduction in India is the increasing accessibility to, and affordability of, internet services. As the internet spreads throughout the population, there are increasing chances of the introduction of a successful telemedicine market in the country. Since, at present, there has been a lack of competition and no big telemedicine giants have penetrated the Indian market, it becomes ideal even for small companies to venture into the Indian market for telemedicine services and to achieve considerable profits. The country is investing in an improvement in its healthcare system and therefore provides a suitable market for telemedicine expansion. Future trends in the telemedicine market signal that there is a potential of about 19% market growth in the industry globally (Bobdey et al., 2019). It can be inferred that developing countries, which are in urgent need of better and technically advanced healthcare services, may play a crucial role in determining the rate of this growth and further direct the future of the global telemedicine industry. There is a high likelihood of telemedicine market expansion and success in countries such as India and, hence, it is appropriate for the business organizations to introduce or collaborate to penetrate this market and achieve the maximum returns.

Ernst and Young (2020) projects that India's telemedicine market will reach more than US$5 billion by the year 2025, impelled by a rise in tele-consultations, tele-pathology, tele-radiology and e-pharmacy as a result of the COVID-19 pandemic. These four sectors account for 95% of the total telemedicine market. Of the four sectors, the market for tele-consultations alone in India is estimated to rise from US$100 million to US$700 million dollars by 2025. This growth will be driven by the regulatory guidelines which were introduced in early 2020. This spending is likely to enable a shift to new consumer behaviour, digital usage and the country's flourishing start-up ecosystem.

7.6 DEVELOPMENT OF A STRATEGIC HAT MODEL TO ILLUSTRATE SUCCESSFUL TELEMEDICINE IMPLEMENTATION IN INDIA

It is also significant to consider that different tele-health services are being accessed at different geographical locations around the globe. Therefore, it is an essential consideration that even if the application of telemedicine in India were to be robust and revolutionary, it should only be introduced after a rigorous analysis of the local interventions and population demands (Mulders, 2019). Therefore, an effective management model must be applied for the successful implementation and establishment of the telemedicine market in India. A management model is defined as the choices made by the establishment authorities to define their objectives, encourage effort, activity coordination, resource allocation and performance of overall management (Van et al., 2017). In the case of the telemedicine market expansion in India, the thinking HAT model of management is suitable. A HAT model helps in the assessment of the current scenario and also directs the information towards the prospects of the market (Van et al., 2017). Hence, the application of the HAT model will not only brief about the present market niche of telemedicine in India but also help in the overall development of the tele-medicine industry. A maturity model also provides insights into "maturity indicators" and provides an opportunity for comprehensive decision-making that can assist in providing a perspective regarding market development and product performance.

The thinking HAT model is composed of different phases of growth and development. These include strategizing, organizing, the consideration of public policy, development and implementation and optimization. Several factors play a role in determining the success of application of this model in the Indian market which can allow for a layered implementation of tele-medicine technology. These factors include (Mulders, 2019):

1. Core readiness or availability of technology: Planning and integration of the model.

2. Technological advancement and readiness: Reliability, availability and affordability of information communications and technology and associated infrastructure.
3. Readiness towards learning: Availability of resources to train the users with the technology.
4. Societal readiness: Collaboration and interaction between institutions to enhance community participation.
5. Policy readiness: Government policies and structures to promote telemedicine via ease of licensing, liability and reimbursements.

The model suggests the implementation of telemedicine by working on a multi-tier system that includes strategic analysis, management and team organization, policy planning, development, execution and optimization of the goods and services (Colombo & Dawid, 2018) (Figure 7.1).

7.6.1 Strategic-level management

Analysis of the market is carried out to obtain a piece of contextual information that helps to determine the priorities and the needs of the population.

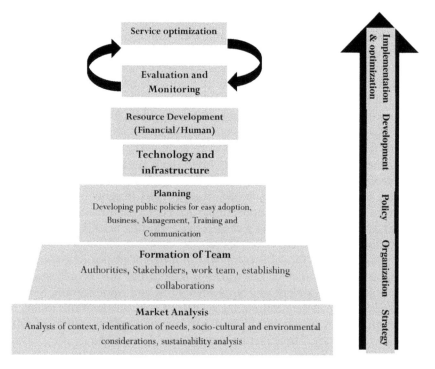

Figure 7.1 Strategic HAT model illustrated for telemedicine implementation in India.

This can assist in the allocation of suitable resources and incorporation of the required assets. Further, evaluation of needs is done by accessing the data and information of the concerned resources and determines whether or not it meets the expectations of the final users of the services (Fesselmeyer et al., 2016). This process is inclusive of information collected about the delivery of the services, organizational issues and general acceptance of the population towards telemedicine. A technological application is successful only if it involves social, cultural, economic and environmental considerations.

7.6.2 Organizational-level management

In the healthcare setting, organizational management must include management at the intra-organizational level, the inter-organizational level and the health system setting. Intra-organizational management is inclusive of informational sharing among the patients regarding telemedicine. Inter-organizational management is inclusive of cooperation and coordination among the healthcare facility providers through networking and the health system setting management is inclusive of the management and development of the health plans (Fesselmeyer et al., 2016). The organizational process, structure, budget, control, and culture must be optimized for the successful implementation of telemedicine in India (Table 7.1).

7.6.3 Policy-level management

It is crucial to develop public policy considerations that are conducive to the working environment and enhance the operational efficiency of telemedicine in India. It is important to encourage population participation by developing and designing integrated public policies for the general masses and the easy adaption of the telemedicine industry in the country (Colombo & Dawid, 2018). The policies must be developed in regard with the local demographics and therefore the premise of the policy must be to direct more profit by working on the needs and demands of the people for successful introduction in the expansion of the telemedicine industry in the Indian market.

7.6.4 Development of telemedicine service

The tele-health service must be developed taking into account the legal, regulatory and security issues that may arise in the industry introduction so that the data is protected and handled responsibly (Colombo & Dawid, 2018). Further, technology development is also inclusive of the development of technological infrastructure, the development of human resource infrastructure and the availability of adequate finances.

Table 7.1 Management considerations in organizational management for the telemedicine market in India

Management considerations	Application
Process	✓ Focus on clear, proactive and identifiable leadership ✓ Establish effective networking and communication mechanisms ✓ Definition of telemedicine in association with healthcare model ✓ Establish health policy objectives ✓ Establish realistic expectations ✓ Analyze internal and external factors that may affect the telemedicine market in India
Structure	✓ Involve health professionals that promote telemedicine ✓ Align telemedicine projects with different participants ✓ Establish training options ✓ Develop telemedicine program guidelines and work load on implementation
Budget	✓ Prepare and implement business plan ✓ Organize required resources for implementation ✓ Establish collaboration ✓ Develop implementation cost estimates
Management and control	✓ Establish strong mechanisms of governance ✓ Control operations and services
Organizational culture	✓ Consider the "resistance to change" factor of the target population ✓ Keep current opinions of telemedicine in considerations for suggestive improvements ✓ Consider different interests, priorities and concerns of the participants ✓ Connect with the needs of participants to match the supply with the demands

7.6.5 Evaluation and optimization

Evaluation and operational enhancement are a continuous process and therefore must be cooperated with and developed regularly (Van et al., 2017). This can be done by supervising the operations of tele-health services so that the tele-health services are feasible and relevant and that they are made available in a timely fashion.

7.7 CONCLUSION

This chapter provides crucial insights by providing concise explanations about the concept of telemedicine. The true potential of the telemedicine market expansion in India has been identified because of the easy access and affordability of the internet services in the country and a huge difference in the level of accessibility of the health services available in the remote and rural regions when compared with the urban areas of India. Further, the doctor–patient ratio in India also highlights the excessive shortage of medical professionals in the region and, therefore, the application of telemedicine in India may aid in the improvement of the country's overall healthcare structure.

The chapter indicates that even when there is substantial scope for the development of the telemedicine market in India as it has quite high receptivity among the healthcare professionals that are willing to develop and advance the telemedicine sector of India. It is inferred that telemedicine in India is highly beneficial and an efficient strategy for medical professionals as it helps tremendously with obtaining specialist opinions, online lab results, image transmission for diagnosis and treatment, monitoring and post-discharge and post-operative care, and video consultations, along with self-care and management. However, the concerns that are associated with the application of telemedicine in the Indian healthcare market were the limited possibility for the replacement of face-to-face consultations with telemedicine, limited internet access and availability, resistance to change, legal responsibilities, and concerns about patient safety and data protection, all of which have hindered the application of tele-health in India. Substantial power cuts and the limitations on internet provision in the rural regions of our country have also been regarded as major barriers in the application of technology in the industry.

Further, a HAT management model provides a comprehensive detailed structure for the establishment and enhancement of India's telemedicine market.

REFERENCES

Abrams, H. B. (2014). What is a health systems innovation centre and why does everyone want one?.*General Medicine*, *15*(1), 5–13.

Agboola, S., Hale, T. M., Masters, C., Kvedar, J., & Jethwani, K. (2014). "Real-world" practical evaluation strategies: A review of telehealth evaluation. *JMIR Research Protocols*, *3*(4), 75.

Ahmed, S. S. T., Sandhya, M., & Shankar, S. (2019). ICT's role in building and understanding Indian telemedicine environment: A study. In *Information and communication technology for competitive strategies*, Editors: Fong, S., Akashe, S., & Mahalle, P. N. (Eds.). (pp. 391–397). Singapore: Springer.

American Telemedicine Association. (2014). What is telemedicine? Retrieved from http://www.americantelemed.org/about-telemedicine/what-is-telemedicine

Amit, R., & Zott, C. (2012). Creating value through business model innovation. *MIT Slogan Management Review*, *53*(3), 41–49.

Armfield, N., Donovan, T., Bensink, M., & Smith, A. (2012). The costs and potential savings of telemedicine for acute care neonatal consultation: Preliminary findings. *Journal of Telemedicine and Telecare*, *18*(8), 429–433.

Baker, L. C., Macaulay, D. S., Sorg, R. A., Diener, M. D., Johnson, S. J., & Birnbaum, H. G. (2013). Effects of care management and telehealth: A longitudinal analysis using medicare data. *Journal of the American Geriatrics Society*, *61*(9), 1560–1567.

Bashshur, R. L., Shannon, G. W., & Smith, B. R. (2014). The empirical foundations of telemedicine interventions for chronic disease management. *Telemedicine Journal And E-Health*, 20(9), 769–780.

Bobdey, S., Narayan, S., Ilankumaran, M., Vishwanath, G., Singh, M. V., Sinha, A. K., & Maramraj, K. (2019). Telemedicine: A force multiplier of combat medical care in the Indian Navy. *Journal of Marine Medical Society*, 21(2), 108.

Bozikov, J. (2015). E-health and m-health: Great potentials for health and wellbeing, but also for harmonization and European integration in health. *South Eastern European Journal of Public Health*, 4(1), 1–7.

Burke, B. L., & Hall, R. W. (2015). Telemedicine: Pediatric applications. *Pediatrics*, 136(1), 293–308.

Cafazzo, J. A., Casselman, M., Hamming, N., Katzman, D. K., & Palmert, M. R. (2012). Design of anmHealth app for the self-management of adolescent type 1 diabetes: A pilot study. *Journal of Medical Internet Research*, 14(3), e70.

Chen, S., Cheng, A., & Mehta, K. (2013). A review of telemedicine business models. *Telemedicine and e-Health*, 19(4), 287–297.

Ciere, Y., Cartwright, M., & Newman, S. (2012). A systematic review of the mediating role of knowledge, self-efficacy, and self-care behaviour in telehealth patients with heart failure. *Journal of Telemedicine & Telecare*, 18(7), 384–391.

Colombo, L., & Dawid, H. (2018). A dynamic model of firms' strategic location choice. In *The economy as a complex spatial system*, Editors: Commendatore, P., Kubin, I., Bougheas, S., Kirman, A., Kopel, M., & Bischi, G. I. (pp. 159–177). USA: Springer, Cham.

Darkins, A., Kendall, S., Edmonson, E., Young, M., & Stressel, P. (2014). Reduced cost and mortality using home telehealth to promote self-management of complex medical conditions: A retrospective matched cohort study of 4,999 veteran patients. *Telemedicine Journal and E-Health. Advance Online Publication*, 31(8), 653–661.

De San Miguel, K., Smith, J., & Lewin, G. (2013). Telehealth remote monitoring for community-dwelling older adults with chronic obstructive pulmonary disease. *Telemedicine and e-Health*, 19(9), 652–657.

Desko, L., & Nazario, M. (2014). Evaluation of a clinical video telehealth pain management clinic. *Journal of Pain & Palliative Care Pharmacotherapy*, 28(4), 359–366.

Doarn, C., Pruitt, S., Jacobs, J., Harris, Y., Vott, D., Riley, W., Oliver, A. (2014). Federal efforts to define and advance telehealth—A work in progress. *Telemedicine Journal and E-Health*, 20(5), 409–418.

Ernst & Young LLP (2020). Healthcare goes mobile: Evolution of teleconsultation and e-pharmacy in new Normal. Retrieved from: https://assets.ey.com/content/dam/ey-sites/ey-com/en_in/topics/health/2020/09/healthcare-goes-mobile-evolution-of-teleconsultation-and-e-pharmacy-in-new-normal.pdf

Fesselmeyer, E., Mirman, L. J., & Santugini, M. (2016). Strategic interactions in a one-sector growth model. *Dynamic Games and Applications*, 6(2), 209–224.

Gellis, Z., Kenaley, B., McGinty, J., Bardelli, E., & Davitt, J. (2012). Outcomes of a telehealth intervention for homebound older adults with heart or chronic respiratory failure: A randomized controlled trial. *Gerontologist*, 52(4), 541–552.

Grabowski, D., & O'Malley, A. (2014). Use of telemedicine can reduce hospitalizations of nursing home residents and generate savings for medicare. *Health Affairs*, 33(2), 244–250.

Hamel, M. B., Cortez, N. G., Cohen, I. G., & Kesselheim, A. S. (2014). FDA regulation of mobile health technologies. *The New England Journal of Medicine*, *371*(4), 372.

Henderson, C., Knapp, M., Fernandez, J. L., Beecham, J., Shashivadan, P. H., Cartwright, M., Newman, S. (2013). Cost-effectiveness of telehealth for patients with long-term conditions (Whole Systems Demonstrator telehealth questionnaire study): Nested economic evaluation in a pragmatic, cluster randomised controlled trial. *British Medical Journal*, *346*, f1035.

Holeman, I., Evans, J., Kane, D., Grant, L., Pagliari, C., & Weller, D. (2014). Mobile health for cancer in low to middle income countries: Priorities for research and development. *European Journal of Cancer Care*, *23*(6), 750–756.

Howitt, P., Darzi, A., Yang, G. Z., Ashrafian, H., Atun, R., Barlow, J., & Cooke, G. S. (2012). Technologies for global health. *The Lancet*, *380*(9840), 507–535.

Istepanaian, R. S., & Zhang, Y. T. (2012). Guest editorial introduction to the special section: 4G health—the long-term evolution of m-health. *IEEE Transactions on Information Technology in Biomedicine*, *16*(1), 1–5.

Johnston, M., Mobasheri, M., King, D., & Darzi, A. (2015). The imperial clarify, design and evaluate (CDE) approach to mHealth app development. *BMJ Innov*, *1*(2), 39–42.

Kayingo, G. (2012). Transforming global health with mobile technologies and social enterprises: Global health and innovation conference. *The Yale Journal of Biology and Medicine*, *85*(3), 425.

Klonoff, D. C. (2013). The current status of mHealth for diabetes: Will it be the next big thing?.*Journal of Diabetes Science and Technology*, *7*(3), 749–758.

Labrique, A. B., Kirk, G. D., Westergaard, R. P., & Merritt, M. W. (2013). Ethical issues in mHealth research involving persons living with HIV/AIDS and substance abuse. *AIDS Research and Treatment*, *203,225*.

Law, L. M., & Wason, J. M. S. (2014). Design of telehealth trials—Introducing adaptive approaches. *International Journal of Medical Informatics*, *83*(12), 870–880.

Lilly, C. M., Motzkus, C., Rincon, T., Cody, S. E., Landry, K., Irwin, R. S., & Group, U. M. C. C. O. (2017). ICU telemedicine program financial outcomes. *Chest*, *151*(2), 286–297.

McGowan, J. J., Cusack, C. M., & Bloomrosen, M. (2012). The future of health IT innovation and informatics: A report from AMIA's 2010 policy meeting. *Journal of the American Medical Informatics Association*, *19*(3), 460–467.

Meher, S. K., Kurwal, N. S., & Suri, A. (2017). E-learning through telemedicine in neurosurgical teaching and patient care. *International Journal of Telemedicine and Clinical Practices*, *2*(1), 2–11.

Mehrotra, A., Jena, A. B., Busch, A. B., Souza, J., Uscher-Pines, L., & Landon, B. E. (2016). Utilization of telemedicine among rural medicare beneficiaries. *JAMA*, *315*(18), 2015–2016.

Mehrotra, A., Paone, S., Martich, G. D., Albert, S. M., & Shevchik, G. J. (2013). A comparison of care at evisits and physician office visits for sinusitis and urinary tract infections. *JAMA Internal Medicine*, *173*(1), 72–74.

Minatodani, D. E., & Berman, S. J. (2013) Home telehealth in high-risk dialysis patients: A 3-year study. *Telemedicine Journal and e-Health*, *19*(7), 520–522.

Mulders, M. (2019). *101 management models*. United Kingdom: Routledge.

Pandya, R., Shaktawat, R. S., & Pandya, N. (2019). ICT enable artificial intelligence in healthcare management in India. In *Computing and network sustainability*, Editors: Peng, S. L., Dey, N., & Bundele, M. (pp. 461–470). Singapore: Springer.

Pearl, R. (2014). Kaiser Permanente Northern California: Current experiences with Internet, mobile, and video technologies. *Health Affairs*, *33*(2), 251–257.

Pratt, S., Bartels, S., Mueser, K. T., Naslund, J. A., Wolfe, R., Pixley, H. S., & Josephson, L. (2013). Feasibility and effectiveness of an automated telehealth intervention to improve illness self-management in people with serious psychiatric and medical disorders. *Psychiatric Rehabilitation Journal*, *36*(4), 297–305.

Rowell, P. D., Pincus, P., White, M., & Smith, A. C. (2014). Telehealth in paediatricorthopaedic surgery in Queensland: A 10-year review. *Australian and New Zealand Journal of Surgery*, *84*(12), 955–959.

Smith, A. (2013). *Technology adoption by lower-income population*. Washington, DC: Pew Research Center. APHSA-ISM Annual Conference. Retrieved from: http://www.pewinternet.org/2013/10/08/technologyadoption-by-lower-income-populations/

Smith, M. W., Hill, M. L., Hopkins, K. L., Kiratli, B. J., & Cronkite, R. C. (2012). A modeled analysis of telehealth methods for treating pressure ulcers after spinal cord injury. *International Journal of Telemedicine and Applications*, *2012*, 729492.

Tamrat, T., & Kachnowski, S. (2012). Special delivery: An analysis of mHealth in maternal and newborn health programs and their outcomes around the world. *Maternal and Child Health Journal*, *16*(5), 1092–1101.

Van Barneveld, T. C., Bhulai, S., & van der Mei, R. D. (2017). A dynamic ambulance management model for rural areas. *Health Care Management Science*, *20*(2), 165–186.

Veit, D., Clemons, E., Benlian, A., Buxmann, P., Hess, T., Kundisch, D., Loos, P., Leimeister, J., & Spann, M. (2014). Business models. *Business & Information Systems Engineering*, *6*(1), 45–53.

Wade, V. A., Eliott, J. A., & Hiller, J. E. (2014). Clinician acceptance is the key factor for sustainable telehealth services. *Qualitative Health Research*, *24*(5), 682–694.

Wakefield, B. J., Holman, J. E., Ray, A., Scherubel, M., Adams, M. R., Hills, S. L., & Rosenthal, G. E. (2012). Outcomes of a home telehealth intervention for patients with diabetes and hypertension. *Telemedicine Journal and e-Health*, *18*(8), 575–579.

Whittaker, F., & Wade, V. (2014). The costs and benefits of technology-enabled home-based cardiac rehabilitation measured in a randomized controlled trial. *Journal of Telemedicine and Telecare*, *20*(7), 419–422.

Wootten, R. (2012). Twenty years of telemedicine in chronic disease management— An evidence synthesis. *Journal of Telemedicine and Telecare*, *18*, 211–220.

Wootton, R., Craig, J., & Patterson, V. (2017). *Introduction to telemedicine*. United Kingdom: CRC Press.

Zickuhr, K., & Smith, A. (2013). *Home broadband 2013 report*. Washington, DC: Pew Research Center. Retrieved from http://www.pewinternet.org/files/oldmedia//Files/Reports/2013/PIP_Broadband%202013_082613.pdf

Chapter 8

The landscape of Big Data and supply chain management

Shweta Sharma

Govt. College for Girls, Sec-14, Gurugram, India

8.1 INTRODUCTION

Big Data is defined as

> a collection of structured (such as historical transactions and financial records), semi-structured (such as web server logs and streaming data from sensors), and unstructured data (such as text, documents, and multimedia files) that is accumulated by businesses and may be mined for information for use in advanced analytics applications like machine learning and predictive modelling.

Data management architectures in organizations now frequently include Big Data processing and storage systems along with technologies that assist Big Data analytics. Big Data is frequently described using the three Vs:

1. Volume: the quantity of data that is appropriate for a range of contexts Stream processing systems, clickstreams, and system logs are a few examples of sources that regularly produce enormous amounts of data.
2. Variety: the enormous range of data kinds that are frequently gathered in Big Data systems. By comparing information from prior sales, refunds, online reviews, and customer support calls, a Big Data analytics project, for instance, can try to estimate product sales.
3. Velocity: The rate or speed at which a lot of data is generated, gathered, and processed. In accordance with the organization's real and non-real time rules, velocity may be monthly, weekly, or yearly.

Recently, several other V's, such as "veracity, value, and variability", have been added to various descriptions of Big Data as shown in Figure 8.1:

DOI: 10.1201/9781003462163-8

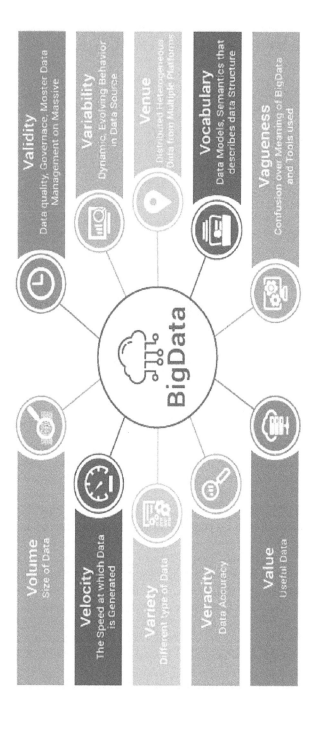

Source: https://miro.medium.com/max/702/1*OYaIBcMGoyYc5IN0ywSXiQ.png

Figure 8.1 Role of various Vs in Big Data.

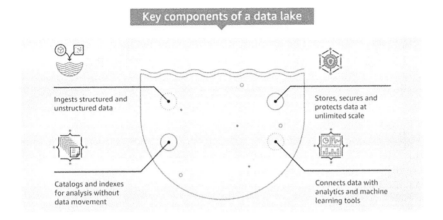

Source: https://www.allthingsdistributed.com/images/datalake2.png

Figure 8.2 Key Components of a Data Lake in Big Data.

Storage and Processing System of Big Data: Big Data is typically kept in the form of a data lake, which is a large pool of data. Big Data is usually kept in a data lake. Data lakes can accommodate a variety of data types and are often built on Hadoop clusters, cloud object storage services, NoSQL databases, or other Big Data platforms, whereas data warehouses are typically based on relational databases and only hold structured data (Figure 8.2).

A distributed architecture is often used in Big Data environments to integrate several systems. For instance, a central data lake may be integrated with other platforms such as relational databases or a data warehouse. Data can be left in its unprocessed state in Big Data systems and subsequently filtered and organized as necessary for particular analytics needs. In other instances, information is pre-processed utilizing software for data preparation and mining to make it suitable for standard applications.

The computational infrastructure is under a lot of demand due to Big Data processing. The necessary computational capacity is usually provided by clustered systems, which divide processing tasks among hundreds or thousands of commodity servers using tools like Hadoop and the Spark processing engine.

It is challenging to get that amount of processing power at an affordable price. Big Data systems are consequently increasingly being housed in the cloud. Organizations have the option of developing their own cloud-based solutions or using managed Big Data as a service from cloud providers. Users of the cloud can increase the number of servers needed to finish Big Data analytics projects. Cloud instances can be disabled until they are once again needed, and the business only pays for the storage and computing time that is used.

Big Data Analytics: The amount of data communicated is enormous in a time when technology has fully taken over and reached its pinnacle.

Massive amounts of data are gathered every day that cannot be processed by a regular computing tool. Big Data is the term used to describe these enormous data sets. The term "Big Data analytics" refers to the process of applying cutting-edge analytical techniques to analyse vast quantities of varied data sets. Organized, semi-structured, and unstructured data from numerous sources, in sizes ranging from terabytes to zettabytes, make up these distinct data sets. They fall under the category of Big Data.

Many different sectors have identified numerous applications for Big Data analytics. It has made it possible for businesses to understand their clients better than they understand themselves, proving the technique's worth.

Putting Big Data Analytics into Practice Data scientists and other data analysts need to have a deep awareness of the available data as well as a clear idea of what they are looking for in it in order to produce trustworthy and pertinent results using Big Data analytics technologies. Data preparation is therefore a crucial first step in the analytics process and involves data profiling, cleansing, validation, and transformation.

Once the data has been collected and made ready for analysis, a variety of data science and advanced analytics disciplines can be deployed to run various applications. These technologies offer Big Data analytics features and capabilities. Text mining, predictive modelling, data mining, statistical analysis, streaming analytics, and machine learning, including its deep learning subfield, are some of these disciplines.

Let's imagine that in order for a business to survive in the cutthroat global economy, we need to understand the customer's behaviour and responses. The following are some of the several types of analytics that huge data sets can be used for:

- Comparative analysis This analyses consumer behaviour analytics and live customer interaction to compare a company's products, services, and branding against those of its competitors.
- Social media surveillance This looks at what customers are saying about a business or product on social media. This can help identify potential problems and the ideal target market for advertising campaigns.
- Marketing analytics and surveys. This provides information that can be used to improve sales promotions and advertising campaigns for products, services, and business endeavours.
- Sentiment analysis Through analysis of the information gathered on customers, it is possible to learn about customer satisfaction levels, opinions toward a company or brand, potential issues, and strategies to improve customer service.

8.2 CLASSIFICATION OF BIG DATA ANALYTICS

Different strategies are needed for the various types of data. The four different forms of big data analytics are the result of this various analytical methodology.

Different strategies are needed for the various types of data. Big data analytics is classified into four subcategories which are:

- Descriptive Analytics
- Diagnostic Analytics
- Predictive Analytics
- Prescriptive Analytics

8.2.1 Descriptive analytics

It is believed that the application of descriptive analytics will enable the identification of descriptive trends within a certain client segment. It simplifies the data while also condensing historical data into a manner that may be easily consumed.

The use of descriptive analytics can help you learn more about past occurrences and trends. This makes it easier to create reports on a business's profits, sales, and other indicators. Examples of descriptive analytics include market basket analysis's use of association rules, grouping, and summary statistics.

8.2.2 Diagnostic analytics

As the name suggests, diagnostic analytics detects a problem. It offers a thorough grasp of the fundamental cause of a problem. Data scientists utilize this analysis want to determine the cause of a particular occurrence. Drill-down, data mining, data recovery, churn reason analysis, and customer health score analysis are examples of diagnostic analytics techniques.

8.2.3 Predictive analytics

As is suggested by the name, predictive analytics is focused with predicting future events. These potential future events can include market trends, consumer trends, and other comparable market-related events. This type of analytics analyses both historical and current data to estimate future events. The most widely used form of analytics in organizations is this one. Both consumers and service providers gain from predictive analytics. It keeps track of our prior deeds and makes predictions about our future deeds in light of them.

Using techniques such as data mining, artificial intelligence, and machine learning, predictive analytics examines the most recent data and predicts what might occur under particular conditions. The "next best offers, churn risk, and renewal risk analysis" are a few instances of predictive analytics.

8.2.4 Prescriptive analytics

The most beneficial but least used sort of analytics is prescriptive analytics. It represents the following stage in the development of predictive analytics. Based on the findings of the descriptive and predictive analytics on a particular data set, the prescriptive analysis examines a number of potential courses of action and offers recommendations. Prescriptive analytics combines business rules and data. Data from prescriptive analytics can be "internal (originating from within the organization) and external" (social media insights).

Two examples of prescriptive analytics for customer retention are the "next best action and the next best offer analysis".

8.3 ADVANTAGES OF BIG DATA ANALYTICS

Big data analytics has benefitted commercial organizations. They use big data analytics in a variety of ways. Due to the advantages it offers, it has grown to become one of the most widely used modern technologies.

  Risk Management: Big Data analytics helps organizations assess their position and progress by providing crucial information on market trends and customer behaviour. Additionally, they can identify and lower any future hazards with the use of statistical analysis methods such as predictive analytics, prescriptive analytics, and others.

  Product Development and Innovations: Big Data Analytics aids firms in making decisions regarding product development and market acceptance. Big data includes product customer reviews. Businesses use this data to evaluate the performance of their product and determine whether to continue selling it moving forward.

  Faster and More Effective Decision-Making: The world has become faster, and so has the decision-making process. Big data analytics has accelerated the decision-making process. Companies no longer have to wait days (or even months) for a response. The speedier response time has also led to greater efficiency. Now, companies can utilize this tactic to modify their business models in the case in which customers are dissatisfied with their goods or services, sparing them from suffering substantial losses.

  Enhance the customer experience: Businesses that routinely monitor customer behaviour may improve the customer experience, even on a personal level. Diagnostic analytics can be used to recognise client issues and locate remedies. In the end, the client will benefit from a more customized experience as a result of this.

Complex Supplier Networks: Businesses may offer supplier networks, also known as B2B communities, with greater precision thanks to big data. Big Data analytics enables providers to get around the limitations they face. It increases providers' success by enabling them to adopt higher degrees of contextual intelligence.

Coordinated and targeted campaigns: Big data can be used by platforms to give customized products to their intended audience. Big data helps organizations perform sophisticated customer trend analysis so they can save money instead of spending it on unsuccessful advertising initiatives. This entails assessing both point-of-sale and internet transactions. As a result, firms are able to develop successful and focused campaigns that satisfy customer expectations and boost brand loyalty.

Principles of Big Data Analytics: There are few fundamental principles of using Big Data Analytics:

Make sure the objective is clear: When it comes to displaying the numerous levels of information flow that affect your organization, big data may be quite descriptive. This, however, is not the main objective. How big data may assist you in achieving certain business objectives should be the first thing you think about. Statistics and big data analytics should result in measurable outcomes, serving as a tool for expanding commercial operations. After all, big data is all about this. You can get insights using traditional business intelligence activities and tools, but you want to take it a step further and use big data to increase your company's ability to make wise business decisions.

Recognize trends: Long-term success for your business depends on your ability to stay informed and observe industry trends. However, creating trends is even better, and big data is excellent at this. For instance, if your business tracks online consumer behaviour, you would need to identify trends quickly and take appropriate action. Big data insights can regularly help businesses spot changes in customer behaviour patterns and prepare for paradigm-shifting events in advance. If this is the case, a business may adjust its sales procedures to fit the new model and gain a considerable competitive edge over competitors who would have to adjust once the trend was already well-established.

Analyze appropriate data: There is a vast amount of data that can be gathered. This is advantageous since it gives you plenty of material to work with, but it does not indicate that you should try to expose it all at once. Businesses, on the other hand, ought to limit their attention to pertinent information. If you just need a small sample to derive reliable conclusions, do not plunge yourself headfirst into a data sea.

Create connections: Big data analytics produces outstanding results when it comes to connecting various business components. Continuous analysis of information clusters should be done in relation to other processes, impact factors, and levers. For instance, despite the fact that marketing is intended to increase sales, it occasionally only gives salespeople useless leads. Each lead may be evaluated using big data analytics to identify which ones result in the greatest results and which ones should be rejected. You can raise the productivity of each division of your business by effectively linking two or more of them.

8.4 BIG DATA CHALLENGES

Big Data has been shown to be quite helpful for companies in the current, highly competitive global period, but there are a number of issues with its use and implementation that also need to be resolved:

1. **Inadequate understanding of big data:** Because of this, businesses struggle to complete their big data projects. Employees might not be aware of the data's purpose, origin, importance, or methods of storage and processing. Data experts might comprehend what is happening, but others might not. For instance, staff members may neglect to maintain crucial data backups if they do not comprehend the necessity of data storage. They could not be using databases to store data properly. As a result, it is challenging to find this crucial information when it is required.

2. Implementing and maintaining massive data systems requires new skills that go beyond what database administrators and engineers that specialize in relational technology typically possess. A managed cloud service can help with these issues, but IT managers must keep a close eye on cloud utilization to prevent costs from getting out of hand. Additionally, moving processing workloads and data sets from on-premises to the cloud is typically a challenging process.

3. **Providing data scientists and analysts with easy access to data:** Managing big data systems presents this issue yet again, particularly in remote settings where a variety of platforms and data sources are present. To help analysts identify pertinent data, data management and analytics teams are increasingly building data catalogues with metadata management and data lineage capabilities. Big data integration is frequently difficult, especially when there are problems with data diversity and velocity.

4. Confusion in selecting the best technology for Big Data analysis and storage is a challenge that businesses commonly run into. They end

up making poor decisions and picking the incorrect technologies. As a result, time, money, effort, and working hours are lost.

5. **A lack of qualified data specialists:** Businesses need qualified data specialists to oversee these cutting-edge technology and Big Data solutions. These professionals will include data scientists, data analysts, and data engineers who are adept at using the technologies and deciphering huge data sets.

6. **Data Protection:** Keeping these vast data sets secure is one of the main challenges of big data. Because they are so busy comprehending, storing, and analysing their data sets, businesses frequently postpone data security. Unsecured data repositories could act as a shelter for malicious hackers, so this is a bad idea.

7. **Data integration from various sources:** Data is gathered from a variety of sources in a business, including social networking pages, ERP software, customer logs, financial data, emails, presentations, and reports created by personnel. It's challenging to combine all of this data to produce reports. Businesses usually disregard the topic. However, analysis, reporting, and business intelligence require faultless data integration.

8.5 BENEFITS OF USING BIG DATA IN SUPPLY CHAIN MANAGEMENT

Big data analytics has significantly improved supply chain management. It solves a number of issues at the tactical, operational, and strategic levels. The supply chain as a whole is impacted by big data. It can be anything from finding ways to improve supplier and manufacturer communication to shortening delivery timeframes.

Using analytics reports, decision-makers may monitor performance and improve operational efficiency. Supply chain analytics is used in conjunction with data-driven decisions to reduce costs and improve service levels.

8.6 APPLICATIONS OF BIG DATA IN SUPPLY CHAIN MANAGEMENT

Consumer Behavior and Usage Patterns: Leading telecom firms are investing more and more in big data analytics in order to study the usage patterns and habits of their customers. Thanks to the information obtained from the analytics report, businesses can significantly increase revenue and subscriber retention.

Improving Inventory Management: Top online retailers and big box retailers with enormous inventories face numerous challenges. Operations managers can use big data analytics to get a minute-by-minute snapshot of their business and identify bottlenecks that stymie supply chain operations. Consumer trends also enable businesses to optimize inventory and promote top-selling products.

Product Quality and Temperature Control: Various industries, such as those involved in the food, agriculture, pharmaceutical, and chemical processing chains, must closely monitor and control specific supply chain elements. Even a slight temperature change of a few degrees might affect a product's quality or even make it completely useless.

Due to a lack of technological support to maintain control, over 30% of temperature-controlled products are destroyed or spoiled before they reach their destination.

Real-time monitoring and order processing: Order fulfilment and traceability must be effective in order to boost business efficiency and satisfy customers. Amazon has fundamentally changed the game with its incredibly quick delivery times, anticipated drop-off time warnings, and minute-by-minute tracking.

Thanks to big data, businesses in different industries may offer comparable experiences to customers and clients. Current shipping data can also help to lower the cost of delivery fleet management by optimizing route deployment, delivery timetables, and item positioning.

Machine maintenance: For organizations, unanticipated issues with machinery brought on by defects, poor maintenance, or antiquated equipment can be quite troublesome. Big data platforms and IoT devices can collaborate to send alerts about any issues or anomalies with equipment. Sensors can be used to keep an eye on production, foresee issues, and notify when routine maintenance is needed to keep the machinery in top functioning order.

8.7 CONCLUSION

Using big data is one of the easiest and most efficient strategies to raise your business' success and profitability rates. It is closely related to business analytics and has the potential to offer valuable insights into all of your organization's key departments, from HR to sales and marketing. A well-executed big data strategy can reduce time to market, hasten product development, streamline operating costs, and assist organizations overcome the difficulties involved in implementing and utilizing it. Big supply chain analytics improve decision-making for all supply chain activities by utilizing data and quantitative methodologies. This creates new data that aids supply chain decision-makers in making better choices, from increasing front-line operations to making strategic decisions like selecting the optimal supply chain operating models.

SUGGESTED READINGS

1. Big Data Analytics in Supply Chain Management: Theory and Applications; Iman Rahimi, M. Ali Ülkü, Amir H. Gandomi, Simon Fong, CRC Press.
2. Supply Chain Management in the Big Data Era; Hing Kai Chan, Nachiappan Subramanian, Muhammad Dan-Asabe Abdulrahman, IGI Global.
3. Supply Chain Analytics: Concepts, Techniques and Applications; Kurt Y Liu.

Chapter 9

The emerging roles of artificial intelligence in chemistry and drug design

Simpi Mehta

DPGITM, Gurgaon, Haryana

9.1 INTRODUCTION

Artificial intelligence (AI) refers to computational tools that can replace human intelligence in the performance of specific tasks. It is a broad science that encompasses computer science, cybernetics, neurophysiology, psychology, and linguistics. The technology is currently advancing at breakneck speed, similar to the exponential growth seen in database technology in the late twentieth century. After decades of rapid development, the meaning of AI has broadened to include artificial neural networks, machine learning (ML), deep learning (DL), and other technologies [1–3]. Deep learning, an important branch of AI, can extract features from massive amounts of data automatically. In addition to that deep learning can find information in images that the human eye cannot recognize [4–6]. Large amounts of biological data in the form of big data, which are present in numerous databases all over the world, serve as the building blocks for ML/DL-based approaches and help in the precise identification of patterns and models that can be used to identify therapeutically active molecules with a great deal less time, workforce, and health investment. In recent years, this subset of machine learning techniques has revolutionized several fields, including computer vision, natural language processing, and game playing [7]. There have been various reviews recently that provide helpful introductions to the field [8–18].

Given its interdisciplinary nature, drug discovery has always been driven by novel developments, whether in the physical sciences (such as organic synthesis) or in the biological sciences (such as genome sequencing or advances in receptor pharmacology), or, more recently, when applying computational approaches to the field. Many terms have been used to describe the use of algorithms in drug discovery, including computer-aided drug design (CADD, which frequently refers to more structure-based approaches); structure-activity relationship (SAR) analysis (which aims to relate changes in chemical structure to changes in activity); chemo informatics (which extends SAR analyses to large compound sets, using different methods, and across different bioactivity classes); and, more recently, ML and AI [19, 20].

DOI: 10.1201/9781003462163-9

Unfavorable absorption, distribution, metabolism, excretion, and toxicity (ADMET) properties are a major contributor to the failure of prospective compounds in the drug development pipeline and to the significant waste of time, money, and human resources throughout the drug discovery process [21].

9.2 DRUG DEVELOPMENT CYCLE

The process that organizations, most frequently pharmaceutical firms, must go through in order to develop their goods before they are made available to the general public is known as drug development. The Food and Drug Administration (FDA) claims that there are five steps in the drug development process:

1. **Discovery and Development:** The discovery and development [22] phases make up the first stage of the generation of novel medicines, such as vaccines [23].
2. **Preclinical Research:** A new molecular entity (NME) must go through a process to determine whether it is safe to use in humans before it is tested on humans. To accomplish this, scientists create techniques for testing these substances in animal models. Pre-clinical research can be divided into two categories: in vivo studies, which entail conducting study on living species; and in vitro studies, which are conducted outside of living things like test tubes. These studies, which are typically modest in size, offer details on the NME's toxicity profile. Researchers evaluate the results of preclinical investigations to determine whether human testing of their product is necessary. At this time, organizations can submit to the FDA an Investigational New Drug (IND) application, which must contain all relevant data, including preclinical data [24–25].
3. **Clinical Research:** Clinical trials can start if the FDA accepts an IND that a sponsor (i.e., the company developing the drug) has submitted. Before a medicine is approved, three clinical phase studies are conducted, each with a distinct goal [26]. After the FDA has given its approval, a fourth phase is carried out, which focuses on the product safety and efficacy profiles. The many stages of clinical research are shown in Table 9.1.
4. **FDA Review:** A New Drug Application (NDA) can be submitted to the FDA for approval after a sponsor has enough proof. According to preclinical and clinical testing, their medicine is safe and effective for the intended use. An NDA includes all important details about how the drug should be used for the target audience. Along with data from clinical research, some of the information in an NDA includes proposed labeling, safety updates, data on drug abuse, patent information,

Table 9.1 Stages of clinical research

Phase	Phase I	Phase II	Phase III	Phase IV
Purpose	Safety and dosage	Efficacy and side effects	Efficacy and monitoring of adverse reactions	Safety and efficacy
Participants	20–100 healthy volunteers or people with the disease/ condition	Up to several hundred people with the disease/ condition	300–3000 volunteers who have the disease or condition	Several thousand volunteers who have the disease/ condition
Length of Study	Several months	Several months to 2 years	1–4 years	Months-Years

data from any research that might have been carried out outside of the United States, information on institutional review board compliance, and usage instructions.

The FDA has six to ten months to determine whether to approve a new medicine after receiving a complete NDA. The FDA collaborates with the sponsor during a procedure known as labeling if the NDA is authorized. An applicant produces or improves the prescribing information for their product through the labeling procedure [27]. Through this procedure, the ideal way to utilize a novel pharmaceutical can be described objectively.

5. **FDA Post-Market Safety Monitoring:** It is exceedingly challenging to compile all the information on the new drug product at the time of approval, even though the drug development process offers crucial information about a medicine's safety and efficacy profile. Therefore, once the product reaches the market, the full picture of the drug safety profile will change over time. In order to address this, the FDA has created a number of programmes, including MedWatch and MedSun, where consumers, medical professionals, and manufacturers may report any issues that might arise with a product that has been given the all-clear [28]. The FDA will examine each situation if issues with any approved drug are discovered and take action to address any safety concerns, such as adding warnings to the usage instructions.

The discovery and development phases make up the first stage of the creation of novel medicine, such as vaccines. When scientists and other specialists become aware of cellular targets engaged in a biological process that are presumed to be malfunctioning and cause disease, the discovery phase typically starts. Drug discovery is a linear, consecutive process that involves several

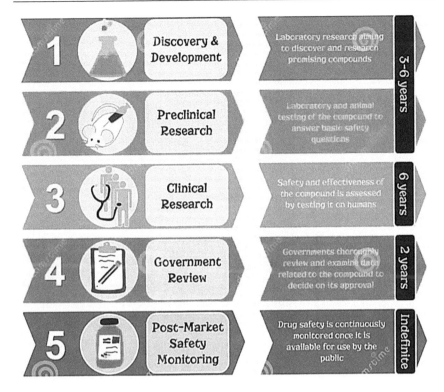

Figure 9.1 Process of drug development.

steps that includes target identification, target validation, assay development, lead identification, lead optimization, preclinical in-vitro and in-vivo studies and clinical phase studies [29]. The new drug development is most tedious, time-consuming and cost-intensive process. For revelation remedial treatment from lead identification through clinical trial is extremely costly and monotonous process and it takes over 14 years. Most of the clinical candidates were discontinued because of poor pharmacokinetics, lack of efficacy, animal toxicity, and severe adverse effects in humans. The efficiency of the compounds with improved ADMET characteristics is then further assessed utilising relevant animal models [30]. Clinical trials are a four-step procedure used to test optimised compounds on human subjects in order to validate and confirm their potency, therapeutic efficacy, ADMET, and potential adverse drug reactions [31]. Each step is carried out in a randomised control manner on a different number of human subjects. If the therapeutic candidate produces the anticipated results during the clinical trial phase, regulatory agencies like the FDA approve it and the drug is then made available on the market. The role of molecular modeling in drug discovery process is represented in Figure 9.2.

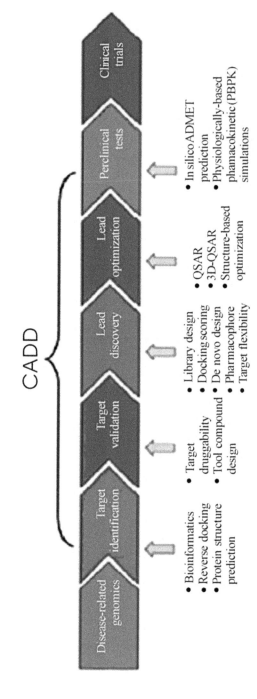

Figure 9.2 Role of molecular modeling in drug discovery process.

9.3 AI IN DRUG DISCOVERY

Given that AI can help with rational drug design [32], decision-making, identifying the best therapy for a patient, including personalised medicines, and managing clinical data for future drug development [33], it is reasonable to expect AI to play a role in pharmaceutical product development from the bench to the bedside. Eularis' E-VAI is an analytical and decision-making AI platform that uses ML algorithms and an intuitive user interface to create analytical roadmaps based on industry rivals, key stakeholders, and current market share to predict key drivers in pharmaceutical sales [34]. This assists marketing executives in allocating resources for maximum market share gain, reversing weak sales, and making better decisions, as demonstrated in Figure 9.3.

A computational model based on the quantitative structure-activity relationship (QSAR) can accurately predict a wide range of compounds or fundamental physicochemical parameters such as log P or log D. These models, however, are far short of predicting complex biological characteristics such as the efficacy and toxicity of substances. Furthermore, QSAR-based models face limited training sets, experimental data errors, and the absence of experimental validations and training sets. Deep learning (DL), a freshly created AI method, can help with these difficulties. The implementation of relevant modelling studies based on big data modelling and evaluation for safety medication effectiveness assessments. Merck participated in a QSAR ML challenge in 2012 to study the benefits of DL in the process of discovering new drugs in the pharmaceutical industry. The open access chemical databases involved include PubChem, ChemBank, Drug-Bank, and ChemDB.

To identify potential therapeutic candidates, QSAR modelling tools, which have evolved into AI-based QSAR techniques such as linear discriminant analysis (LDA), support vector machines (SVMs), random forests (RF), and decision trees, have been used [35–37]. When compared, the ability of six AI algorithms to rank anonymous substances in terms of biological activity to that of traditional methods discovered a statistically insignificant difference [38].

An average of USD 2.8 billion is spent during the more than ten-year drug research and development process. Nevertheless, nine out of every ten medicinal compounds fall short of clinical study Phase II regulatory approval. Methods for predicting in-vivo activity and toxicity for VS include Nearest-Neighbor classifiers, RF, extreme learning machines, SVMs, and deep neural networks (DNNs). Bayer, Roche, and Pfizer are just a few of the biopharmaceutical companies that have collaborated with IT companies to create a platform for drug development for conditions such as immuno-oncology and cardiovascular disease. The numerous VS areas where AI has been applied are outlined below.

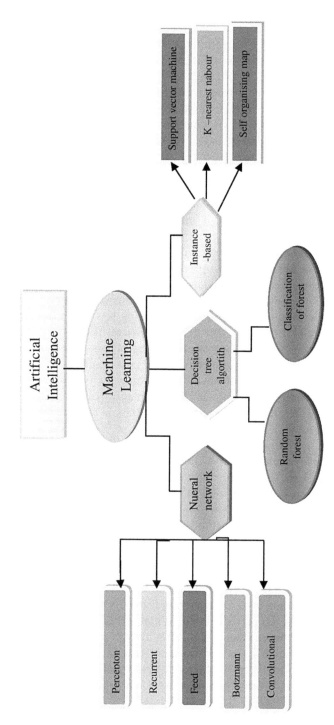

Figure 9.3 Applications of AI in drug discovery and development.

9.4 PHYSIOCHEMICAL PROPERTIES PREDICTION

The physicochemical properties of drugs, such as solubility, partition coefficient (logP), degree of ionisation, and intrinsic permeability, have an indirect impact on their pharmacokinetic properties and target receptor family; as a result, these properties must be considered when developing novel medications. To predict the physicochemical properties of substances, several AI-based techniques can be used. ML, for example, prepares the software using large data sets generated by earlier compound optimization. In drug design algorithms, chemical descriptors such as SMILES strings, potential energy readings, electron density around the molecule, and 3D atomic coordinates are used to build workable molecules and predict their properties.

The Estimation Program Interface (EPI) Suite is a quantitative structure-property relationship (QSPR) process that was developed to determine the six physicochemical properties of environmental chemicals gathered from the Environmental Protection Agency (EPA) [39]. The lipophilicity and solubility of various compounds have been predicted using neural networks built on the ADMET predictor and ALGOPS software [40]. Drug solubility has been predicted using undirected graph recursive neural networks and graph-based convolutional neural networks (CVNN) [41]. To anticipate the acid dissociation constant of substances, ANN-based models, graph kernels, and kernel ridge-based models have all been created in some cases [42]. Similarly, human colon adenocarcinoma (Caco-2) cells and data on the cellular permeability of various compounds were collected using Madin-Darby canine kidney cells and fed into AI-assisted predictors [43]. Six predictive models was developed by using 745 compounds for training to predict the intestinal absorptivity of 497 compounds, including SVMs, ANNs, k-nearest neighbour algorithms, LDAs, probabilistic neural network algorithms, and partial least squares (PLS). These models made use of parameters such as molecular surface area, molecular mass, total hydrogen count, molecular refractivity, molecular volume, logP, total polar surface area, and the polar surface area of the molecule. Similarly, in-silico models based on RF and DNN have been developed to estimate human intestinal absorption of various chemical substances [44]. Thus, AI plays a crucial role in the creation of a medicine by predicting both the desired bioactivity and the intended physicochemical qualities.

9.5 PREDICTION OF BIOACTIVITY

The potency of drugs depends on how well they bind to the target protein or receptor. The therapeutic response cannot be produced by drug molecules that do not bind to or have an affinity for the targeted protein. In a few rare instances, medicinal chemicals may interact with undesirable proteins or receptors and become poisonous. In order to predict drug-target interactions, drug-target binding affinity (DTBA) is crucial. By taking into account the characteristics or similarities between the drug and its target, AI-based methods can determine a drug's binding affinity.

Web programmes such as ChemMapper and the similarity ensemble technique (SEA) are available for predicting drug-target interactions [45]. DTBA has been determined using a variety of methods incorporating ML and DL, including KronRLS, SimBoost, DeepDTA, and PADME. DTBA is calculated using ML-based methods such as Kronecker-regularized least squares (KronRLS), which evaluate the similarity of drug and protein molecules. DTBA was predicted using regression trees, similar to SimBoost, which considers feature-based and similarity-based interactions. Drug characteristics from SMILES, ligand maximal common substructure (LMCS), extended connectivity fingerprint, or a combination of these can also be considered [46]. DL approaches outperform ML approaches because they use network-based techniques that are not dependent on the availability of the 3D protein structure. DeepDTA, PADME, WideDTA, and DeepAffinity are some DL approaches used to evaluate DTBA. The amino acid sequence is entered for the 1D representation of the drug structure and for protein input data. DeepDTA, which accepts drug data in the form of SMILES [46], accepts drug data in the form of SMILES. WideDTA is a CVNN DL approach that determines binding affinity using input data such as protein domains and motifs, amino acid sequences, ligand SMILES (LS), and LMCS [47].

AI-based techniques such as XenoSite, FAME, and SMARTCyp are used to determine drug metabolism sites. Furthermore, tools such as CypRules, MetaSite, MetaPred, SMARTCyp, and WhichCyp were used to identify individual CYP450 isoforms that control the metabolism of a given drug. With high accuracy, SVM-based predictors performed the clearance pathway analysis of 141 approved medicines [48].

9.6 PREDICTION OF TOXICITY

It is essential to predict the toxicity of every pharmaceutical molecule in order to prevent harmful effects. The cost of drug research is increased by the widespread use of cell-based in-vitro testing as exploratory investigations, followed by animal tests to ascertain a compound's toxicity. There are several web-based apps, like Toxtree, LimTox, pkCSM, admetSAR, and others, that can help cut costs. Advanced AI-based approaches investigate chemical similarities based on input variables or estimate the substance's toxicity. Advanced AI-based methods predict a compound's toxicity based on input features or explore for similarity between substances. The Tox21 Data Challenge was organised by the US Food and Drug Administration (FDA), the Environmental Protection Agency (EPA), and the National Institutes of Health (NIH) to test various computational methods for predicting the toxicity of 12707 environmental compounds and drugs [49].

DeepTox, a machine learning algorithm, outperformed all other approaches by identifying static and dynamic features in chemical descriptors of molecules, such as molecular weight (MW).

9.7 RESOURCES USED FOR DEVELOPING DEEP LEARNING APPLICATION

Due to the DL technique's quick evolution, there are now a variety of open source packages and libraries that can be used by both individuals and small groups to investigate the DL without having to create their own DL platform. The majority of these packages provide established GPU computing built-in codes with thorough instructions and annotations. The representative packages are presented with a brief summary and a link in Table 9.2.

Deep learning (DL) holds enormous promise for drug discovery and development, and computer scientists and medicinal chemists have found a common ground to collaborate on the development of DL-based tools, predictive models, and algorithms for drug research and development.

9.8 DRUG DISCOVERY TOOLS

Deep CPI, Deep DTA, Wide DTA, PADME Deep Affinity, and Deep Pocket are public domain DL tools used in the drug discovery paradigm for identifying

Table 9.2 Some packages that are extensively used for practicing deep learning techniques [50]

Package name	Platform/API	Resources
DeepChem	Python	https://deepchem.io/
TensorFlow	Python	https://www.tensorflow.org/
Torch	Lua	http://torch.ch/
Theano	C++/Python	http://deeplearning.net/software/theano/
Caffe	Java	http://caffe.berkeleyvision.org/
DL4J	Python	https://github.com/deeplearning4j/deeplearning4J
Paddle	Python	http://paddlepaddle.org/
Keras	Python	https://keras.io/
CNTK	C++/Python	https://www.microsoft.com/en-us/cognitive-toolkit/
MxNet	R/Python/Julia	http://mxnet.io/
AlexNet	MATLAB	https://www.mathworks.com/products/matlab.html
Pytorch	Python	http://pytorch.org/

drug targets and drug-target interactions. Interactions between bimolecular drugs and chemical substances are known as drug target interactions. Drug target interaction causes the therapeutic effect. Computational approaches can use drug target binding affinity (DTBA) prediction methods and DTI prediction method (Figure 9.4).

AI has a lot of potential to enhance and accelerate drug discovery – the process of identifying potential medicines. The typical drug development processes take around five years to reach the trial stage, but this drug took only a year. The graphic shows a branch diagram for a number of computational strategies for drug target interaction prediction as well as target prediction and identification. These methods are typically categorized as either predictions of drug target interactions or predictions of drug target binding affinities. While DTI prediction tools are classified as docking-based, ligand-based, ML/DL-based gene-based, and text mining-based, DTBA techniques are classified as structure-based and non-structure-based approaches that use ML and DL. Both drug target binding affinity (DTBA) prediction methods and drug target identification (DTI) prediction methods can be used to predict drug targets in reactions. DTI prediction methods include docking simulations, gene ontology-based approaches, ligand-based methods, text mining-based

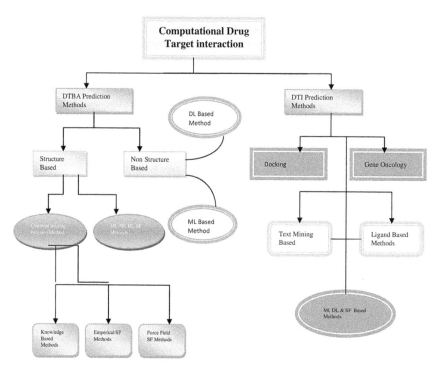

Figure 9.4 Diagram for various computational drug–target interaction predictions.

methods, and ML/DL/Network-based methods. Structure-based approaches are further classified as classical scoring function methods and ML-SF/DL-SF methods, whereas non-structure-based methods are classified as knowledge-based, empirical SF-based, and force-field SF-based methods.

9.9 DTI-CNN

Convolution neural networks (CNN) are a type of neural network (NN) that are commonly used to analyse visual imagery. DTI-CNN [51–54] is a simple DL-based drug-target interaction prediction tool that claims to outperform existing state-of-the-art methods by combining three components: (1) a heterogeneous-network-based feature extractor, (2) a denoising-auto encoder-based feature selector, and (3) a CNN-based interaction predictor. The model can handle low-dimensional feature vectors as well as noisy incomplete and high-dimensional features from heterogeneous data sources such as drug, protein, side-effects, and disease information because it is based on random walk with restart (RWR) and denoising auto encoder (DAE).

9.10 ALPHAFOLD

Proteins, made up of chains of amino acids, are the building blocks of life. AlphaFold [55] has been recognized as a solution for the protein folding structure. In order to gain a better understanding of the structure, the first stage in this contemporary approach entails training a neural network to produce accurate predictions of the distance between residue pairs. Using this knowledge, one can create a mean force potential that can exactly mark determine the protein's structure. Taking the likelihood of less homologous sequences in their sequences, the AlphaFold is a straightforward gradient descent algorithm that aids in optimizing the mean force's potential. Enhancing accuracy without using a complicated sampling procedure [56].

Based on the study of the multiple sequence alignment, AlphaFold often assembles the best probable fragments. By identifying the mutations that have happened over the course of evolutionary timescales in reaction to the other mutations, it analyses the geographic closeness. A significant amount of processing power is required to manage the genuine DNN that discovers evolutionary patterns within protein structure sequences in terms of contact distributions and angular constraints. In addition, instead of a "physically-based reference state," a "learned reference state" is used. Protein-specific statistical potentials can be generated by DL algorithms. This tool has enormous potential for predicting protein structure for scientists [57, 58].

9.11 CRITICAL ASSESSMENT OF PROTEIN STRUCTURE PREDICTION (CASP)

The goal of critical assessment of protein structure prediction (CASP) is to create a technology capable of identifying and generating three-dimensional protein structures from protein sequences [59–61]. Depending on whether or not a template structure was previously available, it can be done in one of two ways: (1) template-based models, or (2) template-free models. If a good template is available, template-based modelling is preferable because it uses the protein structure as the basis for prediction, making it a more advanced technique than template-free modelling and making it easier for researchers to use.

On the other hand, template-free modelling can be used to create the protein structure in the absence of a template. The de novo folding method, one of two approaches for template-free modelling, aims to build three-dimensional structures from scratch using physics fundamentals. Its success is dependent on the efficient application of an accurate energy function to determine the lowest energy state conformation and distinguish native-like structures from decoys. Despite this, fragment-based assembly continues to be dominant due to its accuracy and superior ability to predict protein structures when a suitable template is unavailable.

9.12 CONCLUSION

Innovating and using anticancer drugs has benefited greatly from AI. It is challenging to develop the most effective treatment since humans are constrained by their own degree of understanding. According to this viewpoint, patients who receive the incorrect treatment may miss out on important treatment opportunities and may even see their condition worsen. It can personalise treatment for each cancer patient and provide critical insights and information that cannot be discovered through human identification. AI might accelerate the discovery of novel materials, which would significantly speed up the creation of anticancer medications. Future human cancer research and treatment are anticipated to benefit significantly from AI. In the future, we think AI will significantly improve medical technology.

REFERENCES

[1] Lo, C.M. et al., Cancer quantification from data mining to artificial intelligence. *Comput. Methods Programs Biomed.* (2017) 145:A1.
[2] Abbasi, J., Artificial intelligence tools for sepsis and cancer. *JAMA* (2018) 320(22):2303.

[3] Bi, W.L. et al., Artificial intelligence in cancer imaging: clinical challenges and applications. *CA Cancer J. Clin.* (2019) 69(2):127–157.

[4] Tartar, A. et al., A novel approach to malignant-benign classification of pulmonary nodules by using ensemble learning classifiers. In *Conference Proceedings: Annual International Conference of the IEEE Engineering in Medicine and Biology Society* (2014), pp. 4651–4654.

[5] van der Waal, I., Skin cancer diagnosed using artificial intelligence on clinical images. *Oral Dis.* (2018) 24(6):873–874.

[6] Li, X. et al., Application of artificial intelligence in the diagnosis of multiple primary lung cancer. *Thorac. Cancer* (2019) 10(11):2168–2174.

[7] https://www.theverge.com/2019/3/6/18222203/video-game-ai-future-procedural-generation-deep-learning

[8] Smith, R.G., and Farquhar, A., The road ahead for knowledge management: an AI perspective. *AI Mag.* (2000) 21:17.

[9] Jørgensen, P.B. et al., Deep generative models for molecular. *Sci. Mol. Inf.* (2018) 37:1700133.

[10] Goh, G.B. et al., Deep learning for computational chemistry. *J. Comp. Chem.* (2017) 38:1291–1307.

[11] Jing, Y. et al., Deep learning for drug design: an artificial intelligence paradigm for drug discovery in the big data era. *AAPS J.* (2018) 20:58.

[12] Gawehn, E. et al., Deep learning in drug discovery. *Mol. Inf.* (2016) 35:3–14.

[13] Gawehn, E. et al., Advancing drug discovery via GPU-based deep learning. *Expert Opin. Drug Discov.* (2018) 13:579–582.

[14] Colwell, L.J., Statistical and machine learning approaches to predicting protein–ligand interactions. *Curr. Opin. Struct. Biol.* (2018) 49:123–128.

[15] Zhang, L. et al., From machine learning to deep learning: progress in machine intelligence for rational drug discovery. *Drug Discov. Today* (2017), 22:1680–1685.

[16] Chen, H. et al., The rise of deep learning in drug discovery. *Drug Discov. Today* (2018) 23:1241–1250.

[17] Panteleeva, J. et al., Recent applications of machine learning in medicinal chemistry. *Bioorg. Med. Chem. Lett.* (2018), in press.

[18] Bajorath, J., Data analytics and deep learning in medicinal chemistry. *Future Med. Chem.* (2018) 10:1541–1543.

[19] Lamberti, M.J. et al., A study on the application and use of artificial intelligence to support drug development. *Clin. Ther.* (2019) 41:1414–1426.

[20] Vyas, M. et al., Artificial intelligence: the beginning of a new era in pharmacy profession. *Asian J. Pharm.* (2018) 12:72–76.

[21] Lin, J. et al., The role of absorption, distribution, metabolism, excretion and toxicity in drug discovery. *Curr. Top. Med. Chem.* (2003) 3(10):1125–1154.

[22] The drug development process (2018, January 4). U.S. Food and Drug Administration. https://www.fda.gov/patients/learn-about-drug-and-device-approvals/drug-development-process

[23] Vaccine testing and approval process | CDC. Centers for Disease Control and Prevention. https://www.cdc.gov/vaccines/basics/test-approve.html (2019, April 5).

[24] Vogel, H.G., *Drug discovery and evaluation*, 2nd edition. USA: Springer (2002).

[25] Mohs, R.C., and Greig, N., HDrug discovery and development: role of basic biological research. *Alzheimer's & Dementia* (2017) 3(4):651–657.

[26] Fitzpatrick, S., *The clinical trial protocol.* Buckinghamshire: Institute of Clinical Research (2005).

[27] FDA, *The FDA and the drug development process: how the FDA insures that drugs are safe and effective*, FDA Fact sheet (2002).

[28] Adams, C.P., and Brantner, V.V., *New drug development: estimating entry from human clinical trials.* Bureau of Economics Federal Trade Commission (2003).

[29] Ooms, F., Molecular modeling and computer aided drug design. Examples of their applications in medicinal chemistry. *Curr. Med. Chem.* (2000) 7(2):141–158.

[30] Hooijmans, C.R. et al., Facilitating healthcare decisions by assessing the certainty in the evidence from preclinical animal studies. *PLoS One* (2018) 13:e0187271.

[31] Hefti, F.F., Requirements for a lead compound to become a clinical candidate. *BMC Neurosci.* (2008) 9(Suppl 3):S7.

32] Duch, W. et al., Artificial intelligence approaches for rational drug design and discovery. *Curr. Pharm. Des.* (2007) 13:1497–1508.

[33] Blasiak, A. et al., CURATE.AI: optimizing personalized medicine with artificial intelligence. *SLAS Technol.* (2020) 25:95–105.

[34] Baronzio, G. et al., Overview of methods for overcoming hindrance to drug delivery to tumors, with special attention to tumor interstitial fluid. *Front. Oncol.* (2015) 5:165.

[35] Zhang, L. et al., From machine learning to deep learning: progress in machine intelligence for rational drug discovery. *Drug Discov. Today* (2017) 22:1680–1685.

[36] Jain, N. et al., In silico de novo design of novel NNRTIs: a bio-molecular modelling approach. *RSC Adv.* (2015) 5:14814–14827.

[37] Wang, Y. et al., A comparative study of family-specific protein–ligand complex affinity prediction based on random forest approach. *J. Comput.-Aided Mol. Des.* (2015) 29:349–360.

[38] King, R.D. et al., Comparison of artificial intelligence methods for modeling pharmaceutical QSARS. *Appl. Artif. Intell.* (1995) 9:213–233.

[39] Yang, X. et al., Concepts of artificial intelligence for computer-assisted drug discovery. *Chem. Rev.* (2019) 119:10520–10594.

[40] Lusci, A. et al, Deep architectures and deep learning in chemoinformatics: the prediction of aqueous solubility for drug-like molecules. *J. Chem. Inf. Model* (2013) 53:1563–1575.

[41] Kumar, R. et al., Prediction of human intestinal absorption of compounds using artificial intelligence techniques. *Curr. Drug Discovery Technol.* (2017) 14:244–254.

[42] Rupp, M. et al., Estimation of acid dissociation constants using graph kernels. *Mol. Inf.* (2010) 29:731–740.

[43] Chai, S. et al., A grand product design model for crystallization solvent design. *Comput. Chem. Eng.* (2020) 135:106764.

[44] Thafar, M. et al., Comparison study of computational prediction tools for drug–target binding affinities. *Frontiers Chem.* (2019) 7:1–19.

[45] Lounkine, E. et al., Large-scale prediction and testing of drug activity on side-effect targets. *Nature* (2012) 486:361–367.

[46] Öztürk, H. et al., DeepDTA: deep drug–target binding affinity prediction. *Bioinformatics* (2018) 34:i821–i829.

[47] Gao, K.Y. et al., Interpretable drug target prediction using deep neural representation. In *Proceedings of the Twenty-Seventh International Joint Conference on Artificial Intelligence* (Lang, Je´roˆme, ed.) (2018):3371–3377.

[48] Pu, L. et al., eToxPred: a machine learning-based approach to estimate the toxicity of drug candidates. *BMC Pharmacol. Toxicol.* (2019) 20:2.

[49] Yang, X. et al. Concepts of artificial intelligence for computer-assisted drug discovery. *Chem. Rev.* (2019) 119:10520–10594.

[50] Jing, Y. et al., Deep learning for drug design: an artifcial intelligence paradigm for drug discovery in the big data era. *AAPS J.* (2018) 20(3):58.

[51] Peng, J. et al., A learning-based method for drug-target interaction prediction based on feature representation learning and deep neural network. *BMC Bioinform.* (2020) 21(13):1–13.

[52] Li, Y. et al., Drug–target interaction predication via multi-channel graph neural networks. *Brief Bioinform* (2022) 23(1):bbab346.

[53] Ding, X., and Zhang, B., DeepBAR: a fast and exact method for binding free energy computation. *J. Phys. Chem. Lett.* (2021) 12(10):2509–2515.

[54] Ding, Y., Tang, J., and Guo, F., Identifcation of drug-target interactions via multi-view graph regularized link propagation model. *Neurocomputing* (2021) 461:618–631.

[55] Wei, G.-W., Protein structure prediction beyond AlphaFold. *Nat. Mach. Intell.* (2019) 1(8):336–337.

[56] Jumper, J. et al., Highly accurate protein structure prediction with AlphaFold. *Nature* (2021) 596(7873):583–589.

[57] Ruf, K.M., and Pappu, R.V., AlphaFold and implications for intrinsically disordered proteins. *J. Mol. Biol.* (2021) 433(20):16720.

[58] Batool, M. et al., A structure-based drug discovery paradigm. *Int. J. Mol. Sci.* (2019) 20(11):2783.

[59] Deng, H. et al., Protein structure prediction. *Int. J. Mod. Phys. B* (2018) 32(18):1840009.

[60] Kinch, L.N. et al., Topology evaluation of models for difcult targets in the 14th round of the critical assessment of protein structure prediction (CASP14). *Proteins Struct. Funct. Bioinform.* (2021) 89(12):1673–1686.

[61] Kryshtafovych, A. et al., Critical assessment of methods of protein structure prediction (CASP)—round XIII. *Proteins Struct. Funct. Bioinform.* (2019) 87(12):1011–1020.

Chapter 10

The Impact of the data-driven supply chain on quality

Evidence from the medical device manufacturing industry

Puneeta Ajmera

Delhi Pharmaceutical Sciences & Research University, New Delhi, India

Kartikay Saini

Scottish High International School, Gurugram, Haryana, India

Vineet Jain

Mewat Engineering College, Nuh, India

10.1 INTRODUCTION

The fourth industrial revolution has brought about a number of technologies, from big data to cyber-physical systems, that are essential to maintaining quality and transparency in the operations of supply chains. In this information age, how multidimensional and large data is used to generate insights aids firms in defining and keeping track of supply chain quality (Bag, 2017), as well as how big data analytics plays a role in improving firm performance. Many governments, practitioners, and academic researchers are beginning to accept the idea of supply chain quality management (SC-QM). Supply chain researchers are confident that quality is an essential component of a company's culture and managerial talents and that it can enhance not only the system's operational and financial efficiency but also the performance of the entire supply chain. Total quality management (TQM), quality gurus, quality assurance, quality control, and lean methodologies such as six sigma are the traditional components of an organization's quality management system. A new era of quality management, data-driven SC-QM, can be created through the integration of digital technology with existing methods. Processes are the foundation of healthcare organizations. They must function properly in the field of healthcare, or else someone's life may be in jeopardy. Nurse shortages, staff dissatisfaction, rising expenses, and poor quality are all indications of fundamental issues with the system that the sector is grappling with. As a means of addressing the issues, increasing numbers of healthcare providers are realizing the value of raising quality and safety standards as well as cutting waste. The fact that supply networks

DOI: 10.1201/9781003462163-10

now compete with individual businesses rather than just as autonomous entities is one of the most notable paradigm shifts in contemporary company management. We are now in the era of internet-based competition for business management. It is now suppliers—brand—store vs suppliers—brand—store, or supply—brand—store, as opposed to brand versus brand or store versus store. The ability of management to integrate the company's complex web of commercial contacts will be crucial to the single corporation's long-term success in this competitive environment.

Supply chain management is defined as the planning, structuring, and control of internal and external business processes that enable a chain to produce goods and offer services to customers. Patients, doctors, materials staff, vendors, distributors, and manufacturers all need to collaborate in order to build an effective and efficient clinical chain, even while each party in the chain is competing to turn a profit.

In comparison to the traditional materials management perspective of managing internal, distinct business processes independently, hospitals are beginning to pay attention to the entire supply chain, from suppliers to the delivery of treatment. However, as other industries have discovered in the past ten years, it represents a significant opportunity for cost savings and margin enhancements (Bowersox et al., 2002).

Hospitals, and the healthcare sector as a whole, have not yet devoted much attention to this subject, which is known as supply chain technology. SCM technology is widely employed in the healthcare sector by companies that produce medical supplies and by sizable distributors, but it has not yet made its way into hospitals and to the point of treatment. Supply chain management systems have been widely implemented in numerous industries, including manufacturing, automotive, and retail. In contrast to almost every other industry, hospitals have not significantly implemented the majority of these technologies, and they continue to lag behind in terms of both scope and sophistication.

10.2 THE MEDICAL DEVICE INDUSTRY IN DEVELOPED AND DEVELOPING COUNTRIES

The medical device market is a diverse, forward-thinking, and growing sector. The global market for medical devices is enormous, and it will continue to expand significantly in the coming years. It is anticipated to increase at a Compound Annual Growth Rate (CAGR) of 5.5% between 2022 and 2029, rising from $495.46 billion in 2022 to $718.92 billion in 2029.

The seven segments that make up the majority of the worldwide market for medical devices are given below:

• Diagnostic imaging
• Orthopedic and prosthetic devices

- Patient aids
- Consumables
- Dental products
- IV diagnostics
- Others

With an estimated annual sales of USD 60 billion (INR 3.85 lakh crores) in 2015, diagnostic imaging is the largest segment of the medical device market, accounting for 26% of the total. The second-largest segment, IV diagnostics, has an estimated 24% market share and generates about USD 54.5 billion (INR 3.54 lakh crores). Of the USD 13.68 billion in dental items, dental implants, dental chairs, and dental equipment make up the least portion (INR 88,920 crore). In terms of geography, the Americas (both North and South) are the largest market, accounting for 45% of the world's total medical device sales. Western Europe is the second-largest market (accounting for 27%), followed by Asia (21%).

The pharmaceutical and medical device industry play a significant role in keeping healthcare costs in check and ensuring that the poor have access to it. While still importing almost 70% of its medical devices, India's pharmaceutical industry has become one of the world's top suppliers of generic medications to both developing and industrialized nations over the past decade or so. This poses a significant challenge to providing these equipment to India's impoverished population as well as the poor populations of other developing nations, which is why research on medical device businesses in developing nations is largely focused on this issue.

The design and delivery of safe equipment presents considerable problems for medical device producers. To stay ahead of the competition and utilize emerging technologies such as artificial intelligence (AI), machine learning (ML), augmented reality (AR), the Internet of Things (IoT), and robots, to mention a few, device firms must step up their game in addition to navigating regulations. Since so many technologies are now being used in combination, the new industrial environment has been dubbed Quality 4.0. Industry 4.0, commonly referred to as the fourth industrial revolution, is the subject of Quality 4.0.

10.3 THE HEALTHCARE SUPPLY CHAIN

A traditional healthcare supply chain is a complex network made up of multiple stakeholders at various stages of the value chain. According to Burns (2002), there are three main categories of participants: producers (i.e., companies that make products), purchasers (i.e., wholesalers and distributors), and healthcare providers (hospital systems and integrated delivery networks, or IDNs). Manufacturers make the products, Group purchasing organizations (GPOs) and distributors aggregate many hospitals in an effort to influence economies of scale, and they are funded by administration fees and distribution fees. The products (drugs, equipment, supplies, etc.) are moved, stored, and finally converted into patient-focused healthcare services as part of the healthcare value chain. Many businesses have improved internal skills and quality management procedures to the point where further cost reduction and quality There may not be much added competitive advantage from upgrades. Increasing the supply chain's responsiveness may be the greatest way to reduce expenses, boost output, speed up order processing, and boost revenue.

10.4 BIG DATA

Big data (BD) is an emerging scientific discipline that employs data analysis technologies to improve decision-making. Due to the growing amount of data being collected and stored in these environments, it is currently being largely examined in the e-commerce sector and has significant promise in the healthcare sector. Big data research focuses on finding effective ways to combine and connect massive amounts of data in order to identify patterns, correlations, and other data-related insights. The goal of big data analytics (BDA) is to create valuable insights from massive data, and it is now gaining acceptance. The analytical process, including tool location and use, increases operating efficiency and helps in reducing human errors, which can frequently have disastrous consequences. Contrary to other service industries such as e-commerce and finance, the employment of BDA tools in the healthcare industry is a relatively new occurrence.

10.5 DATA-DRIVEN SUPPLY CHAIN

Industries are under intense pressure to increase the supply chain's (SC's) overall performance in order to outperform competitors. The way that businesses live, breathe, strive, and maintain their competitive advantage in a competitive data-driven environment is through designing and producing data-driven innovations (DDI). Choosing strategies and improving performance has foregrounded the importance of DDI in the strategic competitive performance of the companies, meaning that data is crucial, especially with regard to the retail and e-commerce industries. The adoption of technological solutions such as enterprise resource planning (ERP), radio frequency identification (RFID), the Internet of Things (IoT), etc. has led to an exponential growth in the amount of information or data generated by the many businesses in the present day. As a result, information (data) is growing quickly along with the passage of time.

The exponential growth of data creates problems with an organization's data literacy, data privacy, and data security. In the SC, this is a more pressing issue. Due to the rich and plentiful data in the SC, retail companies must deal with problems such as wasting time on analysing inaccurate and irrelevant data in addition to difficulties with how to store and access enormous amounts of data, etc., which affects the SC's overall performance and efficiency as well as the organizations' profits. The issues faced by data-driven enterprises in the medical device industry are beginning to receive attention. Big data analytics (BDA) can be of great assistance in overcoming this.

The analysis and archiving of the data related to the SC is undertaken done using BDA, which is one of the components of SCA. Big data can be characterized by the following five traits:

- Velocity (how quickly data is created).
- Volume (how much data is produced).
- Value (data should be trustworthy).
- Variety (types of data gathering).
- Veracity (how accurate the data are).

10.6 BIG DATA IN THE MEDICAL DEVICE MANUFACTURING INDUSTRY

Along with continued supply chain difficulties, medical device manufacturers also have to deal with growing but unpredictable demands for specialized just-in-time items, product recall management, and a lack of advanced technological skills. The SC has adopted BDA practices to make it simple to access and retain massive amounts of data. Business intelligence, RFID, IoT, and other techniques can all be applied and incorporated into the SC to give the company a competitive advantage that will enable it to survive. The following are the advantages of Big data in SCM:

1. Fix bottlenecks in the supply chain: For manufacturers of medical devices, a lack of visibility has made the supply chain vulnerable. It is challenging for manufacturers, who frequently have dozens of small distribution locations, to properly plan for production when they cannot view the entire supply chain.

    Manufacturers can receive real-time visibility using advanced analytics, powered by AI and machine learning, which enables resource, fulfilment, and production planning optimization. To meet consumer expectations, manufacturers must be able to modify labour capacity and logistics based on cross-channel inventory levels, raw material availability, and other crucial criteria.

2. More stringent quality control: To make sure that medical equipment satisfies quality requirements, many medical device makers still rely on historical data and routine testing in a real-world production

setting. However, producers become slower and less nimble when they are reactive, which raises prices. The use of AI and machine learning in quality control, in contrast, enables manufacturers to forecast product outcomes using information gathered from across your organization. For instance, a manufacturer may see that batches using a chip from a particular supplier are more likely to experience quality problems than batches using a different chip.

3. Improved employee productivity: Sometimes, due to remote work, and extra sick days, resources are limited. Because of labour scarcities, firms can work harder with less resources through using AI-powered tools. Manufacturers can acquire the data they require to have staff members manage problems quickly and efficiently through investing in sophisticated analytics.

4. Enhance device data collection and utilization: The digital revolution of the medical device sector depends heavily on the Internet of Medical Things (IoMT). With cutting-edge inventions, including smart devices, portable communication devices, and smart sensors, the IoMT has changed medical equipment. This technology has had a significant impact on medical device firms through lowering costs, increasing efficiency, and improving patient outcomes.

5. Right-size inventory levels: The ability of medical device manufacturers to adapt to change depends on efficiency that does not sacrifice quality. They must use technology that minimizes downtime for business-critical assets, and extends the useful life of equipment. To avoid both overstocking and stockouts, they must also appropriately size inventory levels depending on consumer demand and capacity restrictions. Such inventory optimization is essential given the industry's rising demand for personalization. To satisfy demand, it is frequently necessary to produce and stock variations, which is difficult to predict without advanced analysis and forecasting techniques.

10.7 IMPACT OF THE DATA-DRIVEN SUPPLY CHAIN ON QUALITY MANAGEMENT IN THE MEDICAL DEVICE INDUSTRY

10.7.1 Industry 4.0

Before the fourth revolution, the first revolution was sparked by the development of steam power, the manufacture of machines, and the urbanization of farmers. Production was mechanized during the second industrial revolution, and the cost of consumer and industrial goods was reduced through mass production. The third industrial revolution includes the introduction of electronics and control systems, which helped reduce prices while increasing product complexity and lowering expenses. New quality paradigms, methods, and data-driven technologies are being driven by the present-day fourth industrial revolution. Connecting the natural and physical worlds, Industry 4.0 is the beginning of the digital transformation that started with the third revolution. This fourth industrial revolution is powered by the influence of digital data, analytics, connectivity, scalability, and collaboration, which also inform the approaches to Quality 4.0. The democratization of technologies is resulting in transformative capabilities in analytics, material science, and networking as we discover new ways to connect people, devices, and data. Such solutions enable a high-quality change of culture, leadership, collaboration, and standards for the medical device business.

Device designs, functionality, manufacturing procedures, supply chain management, customer service, and procedures for keeping quality systems compliance with regulatory bodies are all being changed by Quality 4.0. Intelligent and linked technologies are being used increasingly frequently as businesses look for a competitive edge to offer innovative items before their rivals.

10.7.2 Quality 4.0

By 2023, the medical device market is expected to reach \$63.43 billion, according to Markets and Markets. The state of healthcare is being improved through the development of smarter, more automated, linked technology that enable remote procedures with medical professionals from across the world. The success of medical device companies on the market in the coming years will depend on their ability to utilize Quality 4.0 technologies. Quality 4.0, which is the use of Industry 4.0 technology to quality initiatives, is seen by LNS Research as following the IoT's lead. According to industry analysts, approximately a quarter of medical device manufacturers use digital transformation technology to boost quality. The design and delivery of items are both done using the same technologies.

To begin with, Quality 4.0 strategies were developed with the aid of digital transformation trends in order to do away with the need for various paper-based quality management systems and procedures. Eliminating manual systems lowers errors, silos, hurdles to collaboration, and traceability problems. Additionally, small and international businesses may grow their design and supply chain processes rapidly thanks to the digitization and automation of production and design processes. RefleXion Medical, a pioneer in biology-guided radiotherapy systems for the treatment of cancer, is one such business that was aware of the necessity of implementing a fully interconnected quality management system (QMS) that could scale to support their path to digital transformation and improved compliance. They needed a platform that was adaptable and could expand along with their workforce, their goods, and their road toward quality compliance. To expedite product development and launches, modern medical device makers rely on distributed teams and supply chains, including design partners, contract manufacturers, and tier-1 component suppliers. The advantages digital transformation technologies may offer to a device manufacturer's product requirements, product capabilities, and regulatory compliance objectives are understood by businesses that have embraced new technology and cloud-based systems.

Digital treatments, medical diagnostic tools, implantable devices, and disposable devices are just a few of the many products in the life sciences industry that aim to be error-free while delivering higher throughput and preserving compliance. Manufacturers of medical equipment and devices face numerous obstacles as they progress from the conceptual and design phases all the way to commercialization. Utilizing Quality 4.0 technology can assist in delivering high-quality and secure devices at every stage of the new product development (NPD) and new introduction (NPI) processes. However, manufacturers are discovering that embracing a newer, higher standard for quality that is led by leadership and supported by everyone involved during the product realization journey is the only way to create designs and quality procedures that are impacted by Quality 4.0. Companies should embrace a more linked, or product-centric, QMS approach in order to satisfy the expectations posed by Quality 4.0 trends. Quality teams need a unified system to spot problems, handle audits, and address quality incidents as product complexity rises due to AI, IoT, robotics, and associated 4.0 technologies. Too many blind spots are created by older, outmoded document-centric QMS techniques. The foundation of a product-centric QMS is the preservation of the entire, intricate product design, which consists of electrical, mechanical, and software components, in a single system. With a direct connection to every element of the underlying product design, this foundation enables comprehensive, integrated quality and corrective action records. Medical device makers can gain from a product-centric QMS in areas including requirements, training, audit readiness, design controls, and integration to upstream and downstream systems.

Additionally, businesses will benefit from having more intelligence-driven insights on their products and quality processes. Better data-driven decisions and cross-functional visibility with the supply chain, engineering, and quality teams will follow from this. With a conventional, document-centric QMS approach, it becomes increasingly challenging to achieve stringent medical device requirements due to the addition of product complexity brought on by these Quality 4.0 transformative technologies.

10.8 CONCLUSION

Professionals and researchers from all around the world are paying attention to the rising costs and operational inefficiency in the healthcare supply chain. A strong supply chain can offer more visibility, better control, and an emphasis on high-quality patient care. Big data analytics (BDA) is currently being adopted gradually with the goal of creating priceless insights from big data. The analytical process, including tool positioning and use, increases operational efficiency and helps in lowering human errors, which are frequently fatal. Big data analytics has the power and promise to revolutionize how those involved in the healthcare supply chain use cutting-edge technology to deliver affordable, timely, and quick healthcare services.

SUGGESTED READINGS

Bag, S. (2017). Big data and predictive analysis is key to superior supply chain performance: a South African experience. *International Journal of Information Systems and Supply Chain Management (IJISSCM)*, 10(2), 66–84.

Bowersox, D. J., Closs, D. J., & Cooper, B. M. (2002). Supply chain logistics management. *Interfaces*, 33(4), 79–81.

Burns, L. R., DeGraaff, R. A., Danzon, P. M., Kimberly, J. R., Kissick, W. L., & Pauly, M. V. (2002). *The Wharton School study of the health care value chain. The health care value chain: producers, purchasers and providers* (pp. 3–26). Jossey-Bass.

Min, H. (2014). *Healthcare supply chain management: basic concepts and principles Business Experts Press.*

Rahimi, I., Gandomi, A. H., Fong, S. J., & Ülkü, M. A. (Eds.). (2020). *Big data analytics in supply chain management: theory and applications*. CRC Press.

Reddy, C. K., & Aggarwal, C. C. (Eds.). (2015). *Healthcare data analytics* (Vol. 36). CRC Press.

Wager, K. A., Lee, F. W., & Glaser, J. P. (2021). *Health care information systems: a practical approach for health care management*. John Wiley & Sons.

Chapter 11

Digital supply chain
Potentials, capabilities and risk management through artificial intelligence

Kartikay Saini
Scottish High International School, Gurugram, Haryana, India

Puneeta Ajmera
Delhi Pharmaceutical Sciences & Research University, New Delhi, India

Vineet Jain
Mewat Engineering College, Nuh, India

Mahesh Chand
J C Bose University of Science & Technology, YMCA, Faridabad, Haryana, India

11.1 INTRODUCTION

Supply chain management (SCM) is the term used for the control of the flow of goods and services, and it encompasses all processes that turn raw materials into completed goods. It involves the purposeful streamlining of a business' supply-side operations to maximize customer value and gain a competitive edge in the market. Through SCM, suppliers try to create and run supply chains that are as efficient and cost-effective as is practicable. Supply chains cover all aspects of production, product development, and the information systems required to coordinate these processes. SCM frequently seeks to centrally link or manage the production, shipping, and distribution of a product. By streamlining the supply chain, businesses can reduce wasteful spending and accelerate the delivery of items to customers. The idea that practically all sold products are the result of the labour of several businesses connected by a supply chain is the basis of SCM. The majority of organizations have only recently realized the value supply chains may add to their operations, despite the fact that they have existed for a very long time. Supply chain management is crucial since it can assist in achieving a number of corporate goals. Controlling production procedures, for example, can enhance product quality while lowering the likelihood of recalls and legal action and assisting in the development of a powerful consumer brand. Controls over shipping processes can also enhance customer service by preventing expensive shortages or periods of inventory overproduction.

11.1.1 Supply chain models

The way supply chain management is implemented varies from company to company. Each business's SCM process is unique due to its specific objectives, limitations, and advantages. A corporation can generally use one of six major models to direct its supply chain management procedures.

The Continuous Flow Model: This is one of the more established supply chain strategies, works best for developed sectors. The continuous flow model assumes that a producer will consistently produce the same good and that customer demand will be relatively stable.

Agile Strategy: This model works best for businesses that provide products which are subject to unexpected fluctuations in demand. This approach places an emphasis on adaptability because a business may have a particular requirement at any given time and must be ready to change course accordingly.

Fast Model: This model places an emphasis on a product's rapid turnover due to its brief life cycle. A corporation uses the rapid chain model to try to take advantage of a trend, produce products quickly, and make sure the product is completely sold before the trend ends.

Flexible Approach: Businesses that are touched by seasonality do well using the flexible model. At the busiest times of the year, certain businesses may have very high demand requirements and very low volume requirements. A flexible supply chain management model ensures that ramping up or shutting down production is simple.

Efficient Model: Businesses that compete in sectors with extremely slim profit margins may try to gain an edge by optimizing their SCM procedures. This entails making the best use of gear and equipment, as well as effectively managing inventories and processing orders.

Custom Model: A corporation can always turn to a custom model if one of the aforementioned models doesn't match its requirements. This is frequently true in the case of highly specialized fields.

Predictable or unanticipated events have always had an impact on supply chains, especially global ones, endangering their profitability and continuity. Therefore, in an effort to lessen the effects of the related hazards, practitioners and researchers have been interested in looking into the origins of these incidents. For three key reasons, this interest has dramatically grown during the past 20 years. First, while the implementation of just-in-time production and logistics practises and lean management may have boosted efficiency, supply chains are still sensitive to unfavourable occurrences because there is little tolerance for mistake and modification. Second, businesses are becoming more globally diversified and less vertically integrated, which makes supply chains more complex and exposes them to more risks. Third, there have been various incidents that have disrupted international supply chains and garnered media attention on a global scale.

11.2 DIGITAL SUPPLY CHAIN

Any machine that uses artificial intelligence to perceive its surroundings and take activities that increase its chances of success in achieving a certain goal is referred to as a digital supply chain. This encompasses a wide range of innovations that let computers solve problems in ways that, at the very least, superficially resemble thinking, such as conventional logic and rules-based systems. As the systems are fed information on the state of a current supply chain situation and then use math, statistics, and heuristics to improve schedules and forecasts, the advanced statistical and optimization techniques that power supply chain planning applications are a type of artificial intelligence. "Digitization" entails a focus on cutting-edge technologies that have the potential to profoundly alter how we do business, particularly supply chain management. Blockchain, additive manufacturing, drones and robotics, artificial intelligence, and machine learning are some of these technologies. These more recent and emerging technologies do not all have the same level of development. But AI is quickly climbing the supply chain maturity curve.

A digital supply chain offers a lot more insight into how the chain is operating. Supply chain owners can build more complicated partnerships with more suppliers because of the improved near real-time visibility of supplier performance and customer needs. As a result, they are protected against the majority of sources of interruptions. Speed, customization, and choice—three today's pillars of excellence in demand fulfillment—are targets of digital supply chains, which are more customer-focused. A digital supply chain blends both organized and unstructured external information with internal systems and data. All stakeholders have complete visibility into the supply chain and may share information with suppliers. Modern technologies gather, track, and analyse data to forecast the future and suggest a course of action in real time (Figure 11.1).

Accelerate Innovation	Reduce Time-to-Market	Maximize Productivity and Efficiency	Reduce Cost	Improve Customer Satisfaction
3x Return on Innovation	20% Reduction in Time-to-Market	30% - 50% Increase Forecast Accuracy	20% - 50% Decrease Scrap and Rework	8% Increase in On-Time Delivery

Figure 11.1 Benefits of digital supply chain.

Source: 1. Farahani, Poorya, Christoph Meier, and Jörg Wilke. "Digital supply chain management agenda for the automotive supplier industry." *Shaping the digital enterprise.* **Springer, Cham, 2017, 157–172. 2. Available from: https://www.netsuite.com/portal/resource/articles/erp/digital-supply-chain.shtml**

11.3 MOST ADVANTAGEOUS SCM DOMAINS FOR AI APPLICATION

Over time, the world has shifted toward a digital future, and Industry 4.0 technologies are now seen as the direction of that future (Kumar et al., 2020). Artificial intelligence (AI) is one of these technologies that has gained the greatest attention (together with blockchain, Internet of Things (IoT), cloud computing, etc.) (Dirican, 2015). According to Deng (2018), AI refers to a machine's ability to mimic and connect with human abilities. When AI is used, problems are solved more quickly, more accurately, and with more inputs. Artificial intelligence (AI), the upcoming technology wave, is already making sense of the flow of operational data coming in from a variety of sensors and cloud apps. Advanced mathematics are being used in this technology to build adaptable, scalable systems, goods, and processes.

When AI is used, problems are solved more quickly, more accurately, and with more inputs. Although AI is not a new topic or academic field of study, technological advancements have only recently revealed that AI has a wide range of applications, garnering attention by adapting processes in a variety of different fields, including SCM. While some information technology fields are becoming competitive necessities, AI technology is starting to emerge as a competitive advantage. In order to increase their functionality, many businesses are transitioning from remote monitoring to control, optimization, and, ultimately, sophisticated autonomous AI-based systems. Along with its growing significance in industry, AI is becoming increasingly prevalent in scholarly discourse. This presence has had an impact on many fields, including business research, which has taken up the subject. AI is now studied from a more comprehensive standpoint, with SCM being identified as one of the industries most likely to benefit from AI applications.

AI applications are significant in three distinct supply chain domains:

1. Product development
2. Procurement
3. Manufacturing

11.3.1 Innovations in product development

Executives in all businesses struggle to deliver the right product to the right consumer at the right time. Additionally, businesses face difficulties integrating capabilities from emerging technologies such as IoT, drones, and robotics with enterprise cloud applications in order to commercialize breakthrough solutions. A lot of research and product development leaders are experimenting with cognitive computing features that are integrated into their operations and products. In addition to assisting with product creation and manufacturing, cognitive computing also supports market research and product development (Figure 11.2).

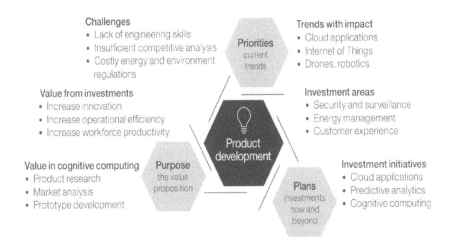

Figure 11.2 Product development priorities, plans and purpose.

Source: IBM Institute for Business Value.

11.4 AI-ASSISTED INTELLIGENT PROCUREMENT

While changing their operational models for revenue growth, procurement officers are concerned with maintaining the health of their global supply chain networks, which are under additional strain from regulations and security issues (Figure 11.3). However, their businesses hold a wealth of information that can be used for deeper procurement insights that is hidden in contracts, transactional systems, and also outside the businesses, among regulators. Learning for supply risk scoring and supplier performance can be enhanced by using cognitive computing capabilities to sort through unstructured data from sources such as news feeds and social networks. These expenditures are expected to increase operational effectiveness, boost revenue, and change the operating model. Cognitive technologies can provide the thorough vision required to obtain real understanding of supply chain risks and interruptions. Procurement professionals can forecast unusual events and make strategies before they happen with this new level of insight.

11.4.1 Manufacturing

Manufacturing executives collaborate closely with their colleagues in product development to market cutting-edge products and advance the growth and profitability goals of their organizations. They are progressing into the automation technologies of adaptive robots and adopting the IoT and cloud's fundamental technologies quickly. And right now, the sector is working on its newest automation innovation, which involves employing AI to make production decisions in real time (Figure 11.4). When a sensor detects

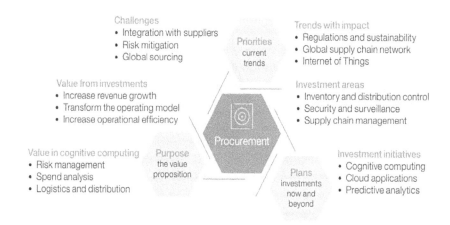

Figure 11.3 Procurement priorities, plans and purpose.

Source: IBM Institute for Business Value.

Challenges
• Skill gaps
• Regulation and traceability
• Product development
  innovation

Priorities
current
trends

Trends with impact
• Cloud applications
• 3D printing
• Internet of Things

Value from investments
• Increase revenue growth
• Increase operational efficiency
• Reduce cost

Investment areas
• Energy management
• Security and surveillance
• Inventory management

Manufacturing

Value in cognitive computing
• Manufacturing/production
• Logistics and distribution
• Quality and preventative
  maintenance

Purpose
the value
proposition

Plans
investments
now and
beyond

Investment initiatives
• Analytics and cloud
• Robotics and IoT
• Cognitive computing

Figure 11.4 Manufacturing priorities, plans and purpose.

Source: IBM Institute for Business Value.

a flaw, for instance, in the manufacturing of automobiles, it sends data to the cloud production application, which immediately demands that the flawed part be taken from the production line and a replacement ordered.

11.5 ADVANTAGES OF AN AI-EMPOWERED DIGITAL SUPPLY CHAIN

11.5.1 Accurate inventory management

The proper flow of goods into and out of a warehouse can be ensured by accurate inventory management. Order processing, picking, and packing are only a few of the many inventory-related variables that can be time-consuming and have a high susceptibility to error. Additionally, effective inventory management can assist in avoiding overstocking, insufficient stock, and unanticipated stock-outs. AI-driven systems can prove to be quite successful in inventory management due to their capacity to handle large amounts of data. Large datasets can be promptly analysed and interpreted by these intelligent systems, providing prompt advice on supply and demand predictions. These AI systems with clever algorithms can forecast seasonal demand and predict and identify new consumer patterns. This use of AI helps reduce the expenses associated with overstocking undesired goods while anticipating future customer demand trends.

11.5.2 Warehouse efficiency

An effective warehouse is a crucial component of the supply chain; automation can help with prompt item retrieval from warehouses and trouble-free delivery to customers. Additionally, AI systems can expeditiously and accurately resolve a variety of warehouse concerns, as well as streamlining complex processes and increase productivity. AI-driven automation initiatives can also drastically cut the demand for, and expense of, warehouse employees in addition to saving important time.

11.5.3 Improved safety

Intelligent planning and effective warehouse management are made possible by AI-based automated technologies, which can improve both worker and material safety. AI may also examine data on workplace safety and alert manufacturers to any potential dangers. It can update operations, record stocking parameters, and perform essential feedback loops and preventative maintenance. This enables manufacturers to respond quickly and forcefully to maintain the safety and compliance of warehouses.

11.5.4 Less expensive operations

The supply chain can greatly benefit from AI systems in this regard. Automated intelligent processes can operate without errors for a longer period of time, decreasing the amount of errors and workplace incidents, from customer service to the warehouse. With their increased speed and accuracy, warehouse robots are more productive.

11.5.5 Prompt delivery

AI systems can assist in reducing reliance on manual efforts, resulting in a faster, safer, and more intelligent overall process. This makes it easier to fulfil the promise of prompt delivery to the consumer. Traditional warehouse processes are accelerated by automated technologies, removing operational bottlenecks along the value chain with the least amount of effort to meet delivery deadlines.

11.5.6 Strengthening of planning and scheduling activities

When faced daily with the experience of increasing globalization, expanded product portfolios, higher complexity, and variable consumer demand, supply chain managers sometimes struggle to build a comprehensive procedure to prepare for successful supply network accounting. This endeavour is made more challenging by the lack of total visibility into current product portfolios resulting from unforeseen occurrences, factory closures, or transportation issues. A typical smart supply chain structure consists of various items, replacement parts, and essential elements that guarantee accurate results. These items or parts can be characterized using a number of attributes that take a variety of values in many supply chain industries. A large range of product variants and applications may come from this. Supply chain managers can improve their decision-making by foreseeing developing bottlenecks, unforeseen abnormalities, and solutions through the implementation of AI in the supply chain and logistics. This will help to streamline production scheduling, which would otherwise be highly variable due to dependencies on manufacturing operations management. Additionally, the use of AI in the supply chain has produced precise forecasts and quantification of anticipated results across various timetable phases, enabling the scheduling of more ideal alternatives as and when such execution hiccups occur.

11.5.7 Intelligent decision-making

AI-led supply chain optimization software which uses intelligent decision-making magnifies crucial choices by utilizing cognitive predictions and suggestions for the best course of action. This could improve the efficiency of the entire supply chain. Additionally, it assists manufacturers with potential time, cost, and revenue implications across a range of scenarios. Additionally, it continuously enhances these recommendations as relative situations change since it learns new things over time.

11.5.8 End-to-end supply chain visibility in AI

Manufacturers must easily have complete visibility of the whole supplier value chain given the intricate web of supply chains that exists today. A single virtualized data layer provided by a cognitive AI-driven automated platform can be used to identify chances for improvement, remove bottleneck procedures, and uncover cause-and-effect relationships. Instead of using redundant historical data, all of this is done using real-time data.

11.5.9 Realistic analytical insights

Many businesses today lack crucial information that may be used to prompt decisions that satisfy expectations quickly and nimbly. In comparison to conventional systems, cognitive automation that harnesses AI's capability can sort through vast amounts of dispersed data to identify patterns and quantify tradeoffs at a scale.

11.5.10 Management of inventory and demand

In order to prevent "stock-out" problems, one of the largest challenges encountered by supply chain firms is to maintain optimal stock levels. Overstocking can also result in excessive storage costs, which, on the other hand, don't generate any revenue. Understanding the art of inventory and warehouse management is what strikes the ideal balance in this situation.

AI and ML concepts produce extremely accurate forecasts of future demand when used for demand forecasting. For instance, it is simple to predict the rise of the market introduction of a new product as well as the decline and end-of-life of a product on a sales channel.

Similar to this, supply chain forecasting using ML and AI ensures that material bills and purchase order data are organized and accurate predictions are created on time. This gives field workers the ability to maintain the optimal levels necessary to meet the demand that is currently (and in the near future).

11.5.11 Improve operating efficiencies

IoT-enabled physical sensors throughout supply chains now provide a gold-mine of information to monitor and modify supply chain planning processes in addition to the gems still mostly buried in disaggregated data system silos at most organizations. Given the billions of sensors and devices available, manually assessing this gold mine can waste a significant amount of operational resources and cause a delay in production cycles. This is where supply chain and logistics with sophisticated analytics enabled by AI offer tremendous value. Radical efficiencies can be attained when supply chain components take on the role of crucial nodes that tap data and power machine learning algorithms. By utilizing machine learning in price planning, the value is realized.

11.5.12 Enterprise Resource Planning (ERP) streamlining

Supply chain managers frequently have more complex business operations than can be handled by standard software packages since they deal with heterogeneous purchasing, procurement, and transportation across worldwide supply chains. The ERP architecture can be streamlined with AI in supply chain and logistics to make it more efficient and intelligently integrate people, processes, and data. Finally, when AI is properly integrated into ERP and associated data systems, data becomes more event-driven over time and receptive, processing larger volumes of data to intelligently learn, quantify, rank, and prescribe fixes more regularly and proactively.

11.6 RISKS AND ISSUES IN THE DIGITAL SUPPLY CHAIN

11.6.1 Security issues

Despite all of its potential, the digital supply chain does carry some risk because it makes use of new technologies and "turns on" shopfloor equipment that were never intended to be online. The potential for Internet of

Things (IoT) security vulnerabilities is one area of study. In the so-called industrial IoT space, where assets and equipment communicate data via sensors and software to enable advancements like predictive maintenance, some of the genuine promise of IoT is found. Sensors can measure a machine's temperature, noise level, or vibratory frequency. Any of these could indicate an impending collapse if they increase. Maintenance crews can address problems before they become critical with accurate monitoring.

This is a crucial strategy for minimizing downtime during the production process. But because this digitalization calls for increased connectivity, risk can spread across the supply chain network and all the way to—or from—the customer. Data corruption or even dangerous machine operation could be caused by bad actors, such as overly zealous rivals or criminals trying to hold processes hostage. Orders may be intercepted or shipments may be redirected.

11.6.2 System complexities

The majority of AI systems are cloud-based, and therefore require a lot of bandwidth to function. Operators may occasionally need specialized hardware to access these AI capabilities, and many supply chain partners may need to make a sizeable initial investment to purchase this technology.

11.6.3 The scalability aspect

The difficulty here is the number of early start-up users or systems required to be more significant and effective, despite the fact that most AI and cloud-based solutions are extremely scalable. This is something that supply chain partners will need to discuss thoroughly with their AI service providers because every AI system is distinct and diverse.

11.6.4 Training costs

Similar to any other new technological solution, training requires substantial time and investment. The supply chain partners will need to collaborate with the AI providers to provide an effective, yet economical training solution throughout the integration phase, which could have an influence on business productivity.

11.6.5 The associated operational expenses

An amazing network of individual processors power an AI-operated machine, and each of these components requires upkeep and repair from time to time. The difficulty in this situation is that the operational investment could be relatively large given the potential cost and energy involved. Manufacturers would also need to replace these, which might significantly increase utility costs and have a negative impact on operating expenses. Capital investments are needed to upgrade a supply chain at a time when many businesses could choose to direct their money elsewhere. Stakeholders in the supply chain may have distinct data needs from manufacturers. For instance, the manufacturer might want throughput data that is so low that it does not allow the store to determine whether or not manufacturing is on schedule.

11.7 DIGITAL SUPPLY CHAIN RISK MANAGEMENT

You must first evaluate your level of digital readiness before making significant investments in new technologies. This assessment is made up of distinct three steps:

11.7.1 Establish realistic expectations

Before committing to the use of AI, every firm must run a self-awareness test. Gather important internal stakeholders and examine the planned implementation's targets and goals with careful inquiries. Decide what you hope to accomplish with the use of new technological integrations if you haven't already had formal discussions about them. Give numbers to your overarching short- and long-term expectations. Compare these to the potential implementation costs, which would include labour costs for installation, setup, and training, as well as the cost of acquiring the necessary technology.

It may be helpful at this point to create new key performance indicators (KPIs) to assess the effects of incorporating AI into supply chain management. These ought to be connected to the long-standing, high-level aims of the organization. Professionals should be aware of the precise contributions that automation and AI would make to various business operations. Even if the deployment is done on a relatively small scale, existing employees and organizational processes will be affected because digital transformation does not happen in a vacuum.

Start considering your project timeframe once you have a sense of: (1) the anticipated return of investment (ROI) of AI; (2) the potential effects of digital transformation; and (3) an estimate of expenditures. Here, long-term gains in efficiency are more important than quick remedies.

11.7.2 Know the technology utilized by the organization

Examine your company's technological readiness after determining what you expect to gain from AI in the supply chain from a more general operational perspective. Three areas should be the focus of that evaluation: people, skills, and tools. To have a better grasp of the potential people effects of technological change, start by speaking with the human resources department. You'll probably need to hire people to fill new positions in your business, meaning that you'll need a strategy for finding and recruiting those candidates. Additionally, you might need to teach current workers to make sure they comprehend how their roles and routines will alter both before and after implementation. Examine your current technology stack and talk to the appropriate stakeholders about its benefits and drawbacks. Try to gauge how well your various technologies are now integrating since interoperability is a crucial indicator of technological preparedness.

Ask why a language is utilized for this application and whether it is also used for other applications to do this. How effective are the instruments for gathering and storing data, and how simple is it to get data when needed? How much do we make use of open-source technologies? Are our mission-critical applications closed, reliant on vendor customization and services, or are they open and ready for application programming interfaces (API)? In conclusion, this assessment necessitates careful planning at the people and application levels as well as broad-based consideration of the state of the entire company.

11.7.3 Thoroughly consider your data

To maximize your profits, you'll need a lot of data because it is the engine that powers AI. Knowing this, the majority of business executives believe they lack the facts necessary to justify an AI investment. This is a typical misunderstanding. A lot of data is typically generated, saved, and forgotten within most organizations. Finding, gathering, and evaluating current data is the challenge these companies are facing, not gathering more data. The majority of data generated by a corporation is frequently used in audits or for compliance-related objectives.

Regardless of the volume, businesses will want to consolidate their business and operational data in order to evaluate their overall data readiness. Additionally, your company probably has more data than you realize. Stakeholders fall prey to a frequent misconception when they argue that there isn't enough data, that it isn't clean, or that they don't know which data is significant. When the underlying problem is availability, they assume scarcity. Most operations could do without siloed data because it isn't useful to them. Organizations may need to invest a lot of time and energy in removing silos before applying AI in supply chain management because these structures are frequently entwined with corporate culture and ingrained business procedures.

Another contributing factor is a lack of communication between various employee types, such as information technology, operations technology, and operations and business. Each of these teams has a unique fundamental goal and approaches data in a unique way. In many organizations, a lack of regular interaction among teams results in a lack of communication regarding crucial items like data. What may be of enormous value to one department is frequently just noise to another.

Internal teams may be forced by digital transitions to break down silos and even reorganize themselves to promote better collaboration. However, in a perfect world, a corporation would eliminate silos before starting a digital transition. By doing this, you'll not only make the transfer process simpler and more efficient, but you'll also learn whether the company is prepared for such a change.

11.7.4 AI integration implementation

The repeated meetings required to discuss solution implementation are a burden that today's supply chain leaders just cannot afford. Actionable insights from integrated AI tools remove bottlenecks and unlock real-time value. This is crucial because supply chain businesses require more action rather than more analysis.

The cost of implementing a complete AI system can vary from millions to tens of millions of dollars, depending on the size of the organization, which makes it seem onerous and prohibitively expensive. Before implementing an analytics programme and integrating AI tools, businesses must fully digitize their operations. Because they don't take into account end-user feedback, businesses frequently waste a lot of resources on this process and wind up going back to fix unanticipated issues.

11.8 FUTURE OF DIGITAL SUPPLY CHAINS

Traditional business models will become stale and eventually outdated as supply chain businesses turn their attention away from products and toward outcomes, leaving behind the bodies and brands of the losers and laggards in their wake. As the foundations of global supply networks deepen, competitive pressures will compel businesses to squeeze every last penny of expense out of their individual operations. This is much more apparent for small, regional, and national businesses that have constrained economies of scale, limited capacity for currency hedging, concentrated markets, and constrained technological and operational budgets. In such situations, examining and adopting the top Software-as-a Service (SaaS) and cloud solutions is a tactic for keeping up with and surpassing the multinational corporations with sizable information technology and operational technology, greater margins of error in the short term, and expensive consultants making poor and expensive supply chain optimization technology mistakes. Companies that are dedicated to supply chain execution excellence, which aims to maximize all available supply chain resources, manage costs at each stage, and deliver products to customers on time and in accordance with specifications, should already be moving in the direction of a digital strategy. There is at present no "Toyota Way" of the digital supply chain that can serve as model for other firms to follow, according to an interesting finding made by McKinsey in its analysis. Businesses may be prone to become bogged down by focusing only on one procedure or measure. The project can also become overly IT-focused, lack a solid business champion, and never advance beyond the pilot stage. Others have trouble getting money and hiring the personnel needed to further digitization.

We are about to witness a paradigm shift from simple reactive intelligence to predictive, adaptive, and continuous learning systems that will drive better decisions for continuous improvements using ML and AI in the supply chain and ML on your existing data sources as all these influences come to bear at the same time.

11.9 CONCLUSION

Automation has had a significant positive impact on the whole supply chain by removing the need for labor-intensive, repetitive jobs. Every automated task that was previously completed by hand now produces fresh data that can be measured and enhanced. In order to improve the digital experiences of their employees and consumers, supply chain organizations should prioritize adopting immersive technologies such as augmented reality (AR) and virtual reality (VR) in the upcoming phase of supply chain optimization. Additionally, all facets of the supply chain will continue to integrate smart technologies like artificial intelligence and machine learning more deeply. By utilizing the data created by automated, connected devices to bridge the information and optimization gaps and better orchestrate supply chains, AI has the potential to enable much more (from sourcing to payments). For those who are willing to develop, adapt, and adopt these technologies, the future presents a supply chain that is far more valuable.

SUGGESTED READINGS

AI in Healthcare: How Artificial Intelligence Is Changing IT Operations and Infrastructure Services 1st Edition; Robert Shimonski. Wiley Publications.

Artificial Intelligence for Healthcare; Cambridge University Press.

Deng, L. (2018). Artificial intelligence in the rising wave of deep learning: The historical path and future outlook [perspectives]. *IEEE Signal Processing Magazine*, 35(1), 180. DOI: 10.1109/MSP.2017.2762725

Dirican, C. (2015). The impacts of robotics, artificial intelligence on business and economics. *Procedia-Social and Behavioral Sciences*, 195, 564–573.

Farahani, Poorya, Christoph Meier, and Jörg Wilke. "Digital supply chain management agenda for the automotive supplier industry." *Shaping the digital enterprise*. Springer, Cham, 2017, 157–172.

Intelligent Healthcare: Applications of AI in eHealth (EAI/Springer Innovations in Communication and Computing).

Machine Learning and AI For Healthcare Big Data For Improved Health Outcomes 2Nd Edn by Panesar, Springer India.

Kumar, R., Singh, R. K., & Dwivedi, Y. K. (2020). Application of industry 4.0 technologies in SMEs for ethical and sustainable operations: Analysis of challenges. *Journal of cleaner production*, 275, 124063.

Chapter 12

Intelligent location specific crop recommendation system using big data analytics framework

J. Madhuri

Bangalore Institute of Technology, Bangalore, India

M. Indiramma and N. Nagarathna

B.M.S. College of Engineering, Bangalore, India

12.1 INTRODUCTION

Unprecedented population growth and related socioeconomic challenges are currently associated with projected future global food shortages [1]. It is estimated that by the year 2050, the world population may increase by 30%, which calls for an acceleration of food productivity by 70%. We need to address the necessity for agriculture to reduce its environmental footprint while adapting to mitigate the effects of a changing climate in order to feed a growing global population. Since the early 2000s the necessity to harness increasing food demands has initiated several studies in the field of agriculture [2, 3]. Hence, there is a need to increase improvements in the field of agriculture taking into account the different factors affecting food production.

Interaction with technology in the agricultural field plays a crucial role in providing affordable and nutritious food to humankind with the looming challenges of climate change [4]. Although there has unquestionably been progress in the use of technology in farming practices, it fails to deliver the equivalent impact globally. Most agricultural innovations fail to reach the intended end users, thereby failing to generate the anticipated improvements in the farming sectors. It is necessary to ensure that the measures should be able to reach more target users and provide support to millions of farmers to improve productivity around the world. Innovations in the agricultural sector have to be driven by the challenges of climate extremities, variable soil conditions, product demand, and the well-being of the end-users such as farmers and consumers. Technology can be conceptualized as a driver for successful innovation. One important step in agricultural innovation would be the support for site-specific crop management strategies based on observing multiple parameters such as climate, soil conditions, and developmental stages of plant and crop health. Precision agriculture leverages the technologies to analyze historical climate patterns, sensor data, and soil variability.

DOI: 10.1201/9781003462163-12

199

The advent of precision agriculture has reformed traditional agriculture by combining Information and Communication Technologies (ICTs), and this has exhibited significant progress in terms of both agricultural throughput and sustainability [5]. Concerning this, agriculture systems utilize inputs from remote sensing and Global Information Systems (GIS), physical measurements of the soil, meteorological data and these all serve as potential inputs to provide a factual recommendation about crop management to the farmers.

The enormous amount of data being generated from various heterogeneous sources indicates the need for the large-scale collection, storage, preprocessing, modeling, and analysis. Compared with traditional data sets, massive unstructured data requires Big Data platforms in order to enable real-time data analysis. This helps us to understand the hidden values in the data, thereby encouraging us to take up new challenges e.g., how to effectually manage and analyze huge data sets [6]. Although Big Data analytics has for some time been a popular term in various industries, such as banking, insurance, retail, and social networking, it is only recently being introduced into the field of agriculture [7]. Our goal in this chapter is to extend our knowledge about harnessing Big Data in the agricultural sector and the expected changes that are caused by Big Data developments. Here, we are focusing on a case study of developing the recommendation model to enable the farmers to choose suitable crops. The recommendation model considers the location-specific climate data, soil data, crop characteristics, and also the pricing trend of the crop. The model makes use of the Apache Spark platform for Big Data analysis.

12.1.1 Potential of big data analytics

Big Data analytics is the term used to explain all the procedures and technologies required for knowledge discovery, including data extraction, loading, transformation, and analysis. It also includes a familiarity with particular tools, methodologies, and approaches to deliver results to the end users and decision-makers [9]. Big Data analytics offers an opportunity to expand standard information extraction methodologies into new fields. This opportunity prompted researchers and technology vendors to create sophisticated platforms, frameworks, and algorithms in order to address the Big Data challenges.

12.1.2 Big data analytics framework

The Big Data analytics framework can be divided into four segments, namely: Data collection, information extraction, data analysis, and data interpretation. The framework is as shown in Figure 12.1.

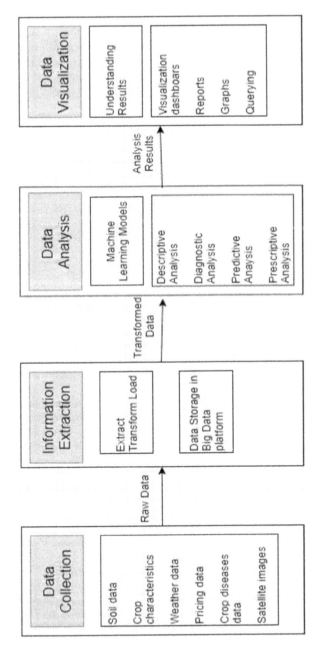

Figure 12.1 Big data analytics framework.

12.1.3 Data collection

The primary phase in data analytics is considered to be data collection. This is the process of the collection, measurements, and evaluation of correct insights for study using approved procedures. Data collection is usually the first and most significant phase in the research process regardless of the subject of study. In agriculture, both structured and unstructured data are gathered from a variety of sources, including weather reports, soil conditions, satellite photos, and more. Open-source datasets available for agriculture analysis are listed in Table 12.1.

12.1.4 Information extraction

The information-extraction phase extracts only the data needed for analysis. Data is interpreted, integrated, mined, analyzed, cleansed, and stored in a data warehouse to obtain information [15].

Table 12.1 Open source data sets

Dataset	Description
Open Government Data (OGD) Platform India [10]	This portal is a single point of access to information, documents, services, tools, and applications authorized by the Government of India's ministries, departments, and organizations.
World Bank open data [8]	This maintains two million data resources and also millions of Database Analyses, which include a collection of time series data. It includes data about agricultural and rural development indicators across the globe.
Online agricultural marketing information system [11]	This maintains pricing information about the selling prices of cereals, vegetables, and fruits in India.
Bhoomi geo portal [12]	This contains information on the soil and site characteristics of India. Soil maps of 1:10000 scales depict the potential area of crops and cropping patterns.
India Meteorological Department [13]	All-India weather forecast data.
Arkansas Plant Disease Database [14]	Pathological (infectious) and non-pathological (physiological/environmental) diseases of agronomic row crops and horticultural crops are shown in these photos. Growers of the crops listed will find these images useful in identifying diseases.

12.1.5 Data analysis

Data analysis is used to help people make better decisions. As agricultural data is both structured and unstructured, Big Data analytics helps in finding trends and analytics to utilize this information and gain farming insights [16].

12.2 AN EFFECTIVE BIG DATA-BASED CROP RECOMMENDATION SYSTEM

The main goal of this work is to construct a recommendation system with an efficient data analytic framework via which the farmer community can receive location-specific crop information. The suggested framework creates a recommendation system that considers soil physical qualities, climatic variables, and crop attributes to identify a suitable crop. Crop yields may be increased by selecting the correct crop to suit the conditions of each location. The recommendation system provides the options to choose the right crop for the farmers. It also assists government agencies in devising appropriate land management methods to boost production and preserve soil fertility. The data pipeline of the recommendation system includes those actions that take the data from its sources, preprocess the data, build the machine learning model and reach the destination for storing the results. To build a framework that works in real time, the proposed system operates on Big Data platforms, namely Apache Kafka and Apache Spark. It uses Elasticsearch to handle the data pipeline efficiently.

12.2.1 Background work

12.2.1.1 Agricultural recommendation system

One of the difficulties faced by Indian farmers is that they do not select the appropriate crop for their land. This will result in a significant drop in production. Precision agriculture is the strategy used to solve the farming issues related to productivity and management. Precision agriculture is a contemporary agricultural approach combining data associated to soil properties, soil types, and crop production statistics in order to recommend the best crop to farmers based on their unique site conditions [17]. This minimizes the chances of crops being chosen incorrectly, and will accordingly contribute to an increase in crop production. One study addresses this challenge by offering a recommendation system based on a machine learning ensemble model [18]. The Internet Of Things (IoT) gives the added option of agriculture to procure farming data at regular intervals globally. Another researcher proposes an irrigation recommendation system based on IoT where the sensors installed on the farm collect the details of the soil, such as temperature, humidity, soil moisture, and wind speed [19]. Based on these inputs and weather parameters, the system recommends five types of irrigation alerts. The suitability of the crops is recommended using weather parameters [20].

12.2.2 Big data platforms used

12.2.2.1 Apache Kafka

Apache Kafka is a high-performance, long-lasting and scalable, publish-subscribe messaging system. It is applied to design real-time data pipelines and streaming applications. Real-time streaming is used in a wide range of businesses and organizations for a number of different purposes.

The major advantage of Kafka is its capacity to receive huge amounts of real-time data with minimal latency, however, staying fault-resilient and scalable. Kafka supports a high number of data streams for reading and writing [22]. Its data streams are securely stored in a cluster that is distributed, duplicated, and resilient to fault. A replication factor of three is a standard production option, which means that your data will always be duplicated three times. A producer sends messages to a Kafka topic (messaging queue). A topic can alternatively be thought of as a message category or a name for the feed where the messages are published.

Producer Application Programming Interface(API) and Consumer API are the two major libraries of Kafka. In practice, an application can use the **Producer API** to transmit a stream of records to Kafka topics. An application can use the **Consumer API** to subscribe to Kafka topics and process the stream of records. A Kafka stream API has higher-level processing features for real-time event streams.

12.2.2.2 Apache Spark

Apache Spark is a distributed open-source processing solution for large data workloads. Data reuse is enabled by the generation of data frames, which is the extension of the Resilient Distributed Dataset (RDD)[21–23]. Apache Spark is currently in the limelight as it has enabled iterative machine learning jobs, interactive data analysis, and batch data analysis. The original design goal of Spark, which stands for "lightning-fast cluster computing platform," was to make data analysis faster and also to be able to develop programs quickly and easily [26]. Spark can also be used for implementing distributed SQL, establishing data pipelines, importing data into a database, running machine learning algorithms, dealing with graphs or data streams, and much more. Spark can deal with both organized and unstructured data, such as CSV files and JSON files.

The components of spark include Spark SQL, Spark Streaming, MLlib, and GraphX. **Spark Streaming** is a Spark library that takes streaming data from Kafka and analyses it in real time.

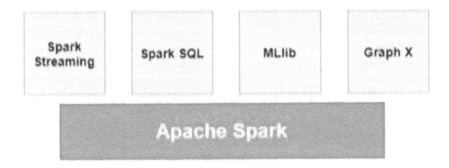

Figure 12.2 Apache Spark software stack.

MLlib is an efficient machine learning library which includes classification, regression, clustering, collaborative filtering, dimensionality reduction, and underlying optimization primitives, as well as common learning algorithms and utilities. **Spark SQL** is a structured data-processing Spark module. It provides DataFrames as a programming abstraction and may also serve as a distributed SQL query engine. Spark SQL can be used to execute SQL queries. **GraphX** is a new element of Spark for graphs and graph-parallel processing. GraphX adds a new Graph abstraction to the Spark RDD: it is a directed multigraph with attributes linked to each vertex and edge. Spark Core is the platform's fundamental general execution engine, on which all other features are built.

12.2.2.3 Apache HBase

HBase is a Hadoop Distributed File System (HDFS)-based column-oriented non-relational database management system. HBase is a fault-tolerant storage system for sparse data sets, which is popular in many Big Data applications [27]. The HBase system is built to scale linearly. It is similar to a typical database in that it consists of a collection of standard tables with rows and columns. The tables in HBase are ordered by row, and it is a column-oriented database. Only column families, or key-value pairs, are defined in the table schema. A table can include several column families. Each column family can contain any number of columns. The values of subsequent columns are saved on the disc in a logical order.

12.2.2.4 Elasticsearch

Elasticsearch is a real-time full-text search and analytics open-source distributed database system [25]. It' is written in Java and has grown in popularity over the years, currently being widely utilized in many common or lesser-known search and data analysis uses of the Apache Lucene library. Elasticsearch receives raw data from several sources, such as logs, system metrics, and web applications. Users can conduct complex searches against their data and use aggregations to receive elaborate summaries of their data once it has been indexed in Elasticsearch. The following are some well-known websites or apps that use Elasticsearch: Stack Overflow (for queries based on geolocation), Wikipedia (for full-text search and suggestions), and GitHub (code searching) [28].

12.2.3 Proposed framework

The proposed framework for the recommendation system is shown in Figure 12.3. It consists of four parts: data producer-consumer model, ML model, data storage, and User Interface (UI) for displaying results.

The UI for the crop recommendation system consists of the input form as shown in Figure 12.4 and crop suitability results are as shown in Figure 12.5.

12.2.3.1 Data collection

In the proposed work, we focus on data sets obtained from the Doddaballapur (dist.) located in Karnataka, India. The district is located at latitude 13° 20′ North and longitude 77° 31′ East. The four crops considered for the recommendation system are maize, finger millet, sugarcane, and rice. The data includes location-specific soil and climatic features. Additional environmental factors, namely precipitation, humidity, wind speed, sunlight hours, and Potential Evaporation Transpiration (PET), are considered in addition to typical meteorological parameters such as rainfall and temperature. Daily meteorological data from 2007 to 2017, for the site, was collected from the open government data platform, in India [10]. The land/soil characteristics for this location are obtained from the National Bureau of Soil Survey and Soil Usage Planning (NBSS & LUP), Bengaluru [12] (Table 12.2).

The data set contains 17 soil physical property measurements and crop characteristics. Six qualities that are common and important for crops have been evaluated among these measures. Texture, soil pH, gravel code, erosion code, and water retention qualities such as slope and depth are all part of the soil data. The soil characteristics are critical for crop development. Soil efficiency has a direct impact on crop development, regardless of nutrient levels [18]. For each crop under investigation, the crop data include mean temperature, soil drainage, texture, depth, slope, and length of growth period [29].

Four sets of crop data for rice, maize, sugarcane, and finger millet are considered as the experimental data set. Categorical data makes up the majority of the data set. One-hot encoding is used to transform categorical data into numerical form, which enables the implementation of machine learning algorithms easier.

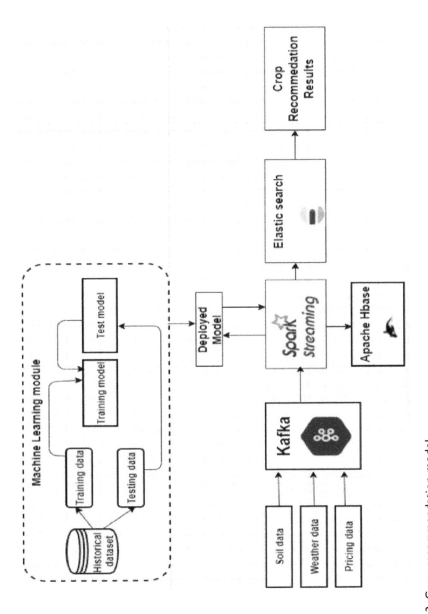

Figure 12.3 Crop recommendation model.

Figure 12.4 Input form.

Reference

Class 1 : Highly Suitable

Class 2 : Moderately Suitable

Class 3 : Marginally Suitable

Class 4 : Not Suitable

Predictions by Neural Network are

Maize can be grown with class 1

Rice can be grown with class 2

Fingermillets can be grown with class 2

Sugarcane can be grown with class 1

Figure 12.5 Crop suitability output.

Table 12.2 Attributes used in the recommendation system

	Dataset parameter	Description
1	Minimum Temperature	Minimum temperature of the day.
2	Maximum Temperature	Maximum temperature of the day.
3	Sunshine hours	The measure of daytime duration
4	Potential evapotranspiration	Calculated from Thornthwaite [29] Quantity of evaporation happening in the region.
5	Soil texture	The texture of the soil 1. Clayey 2. Loamy sand 3. Sandy loam 4. Sandy clay loam 5. Sandy
6	Soil pH	Acidity content measured in soil
7	Gravel code	Indicates particle size in the soil 1. Non-gravel if <15% of soil particles are coarse 2. Gravelly if 15–35% of soil particles are coarse)
8	Erosion code	Amount of topsoil carried away due to wind and water 1. Severe erosion 2. Moderate erosion 3. Slight erosion
9	Slope	Specifies the inclination of the soil surface relative to horizontal. It is indicated by the values 1–4. 1. Level to nearly level if the inclination is 0–1% 2. Very gently sloping if the inclination is 1–3% 3. Gently sloping if the inclination is 3–5% 4. Moderately sloping if the inclination is 5-10%
10	Depth	Specifies the support given by soil for shedding roots and absorption of water. It is indicated by the values 1–3. 1. Shallow if the depth is 25–50 cm 2. Deep if the depth is 100–150 cm 3. Very deep if the depth is >150 cm
11	Mean temperature	The average temperature necessary for the crop
12	Soil drainage	Measures how excess water moves across, through, or out of the soil
13	Soil Potassium	Potassium present in the soil
14	Soil Nitrogen	Nitrogen present in the soil
15	Soil Phosphorus	Phosphorus present in the soil
16	Length of the growing period	Time taken by the crop to sprout and reach the harvest stage
17	Effective root depth	Distance of roots of a crop from the surface down into the ground

12.2.3.2 Data ingestion

The **Apache Kafka** producer acquires the data from different data sources. Our proposed system consists of three topics: soil, weather and prices. Messages under the same topic are stored under different partitions. The Kafka cluster is made up of several brokers, which are server nodes and the partitions of each topic are divided over these brokers. The topics from these partitions are subscribed to by the consumers. The producer is responsible for ensuring that the data is full and consistent until it is viewed by the consumer. The consumers of Kafka streaming data are Spark streaming and HBase.

12.2.3.3 Machine learning model

Spark streaming creates the data frames of the real-time data ingested from Apache Kafka. Apache Spark consists of the machine learning library, MLlib, which runs on the data frames to provide the predictions [21]. The machine learning model was built offline using the historical data of soil, weather, and pricing and is deployed to predict recommendations regarding crop suitability. The algorithms used to build the model are Artificial Neural Networks (ArNN) and decision trees. The pyspark.mllib package facilitates running the deployed model on the streaming data frames. The machine learning model consists of four stages, namely: 1) Data preprocessing, 2) Feature Selection, 3) Machine learning algorithms, and 4) Evaluating metrics for the results

1. Preprocessing is one of the most important phases to efficiently express data for the machine learning algorithm that has to be trained and tested. In our model, the missing values are identified and they are filled by the mean value of that attribute.
2. Feature selection is used to determine the important features in the data set. In the current paper, we are using a hybrid filter and wrapper method for feature selection [28]. The Pearson Correlation coefficient-based filter method is combined with the backward elimination wrapper method to specify important features required for crop recommendation.
3. Machine learning algorithms

The classification algorithms used in this proposed system are

- Multilayer perceptrons (MLP) are fully connected Artificial Neural Networks (ArNN). Neural networks learn from the representation of the training data and relate it to the output variable. It mainly consists of three layers: the input layer, the hidden layers, and the output layer. Neural networks are trained based on their network topology, activation functions, and weight adjustments. The arrangement of neurons with connecting links and associated weights is the network topology. Activation functions that are applied along with the input define the output of that layer and avoid the linearity of the output. In ArNN, learning indicates the adjustments of weights associated with each link. A gradient descent algorithm is used to train the MLP. The following algorithm summarizes the gradient descent

1. Each training example is a pair of the form(x,t), where x is the vector of input values, t is the output values, η is the learning rate
2. Initialize each weight w_i to a random value
3. Until the termination condition is met, Do,
   a.   Initialize each Δw_i to zero.
   b.   For each (x,t) in training examples, Do,
       i.   Input instance x_i to the unit and compute the output o
       ii.   If calculated $o \neq t$, For each linear weight wi, Do

$$\Delta w_i \leftarrow \Delta w_i + \eta(t-o)x_i$$

   c.   For every linear unit Do,

$$\Delta \leftarrow w_i + \Delta w_i$$

Where η is the learning rate (0.005), and w_i is the weight across the connecting links between the layers. We have extended the gradient descent with optimization algorithms to find the optimal values for the neural network parameters [31].

12.2.3.4 Adam optimizer

To provide a more optimized gradient descent, the Adam Optimizer is used. When the global minima are reached, the gradient descent rate is controlled to minimize oscillations while taking sufficient steps to overcome the hurdles of local minima. The momentum and velocity equations are shown in Equation 12.1.

$$m_t = \beta_1 m_{t-1} + (1-\beta_1)\left[\frac{\delta L}{\delta W}\right] \quad v_t = \beta_2 v_t + (1-\beta_2)\left[\frac{\delta L}{\delta w_t}\right]^2 \qquad (12.1)$$

The Adam Optimizer corrects the problem of high oscillations while reaching global minima by computing bias-corrected m_t and v_t. Instead of normal weight parameters, we consider the bias corrected weight parameters, $\bar{m}_t$ and $\bar{v}_t$ The general equation can be shown as in Equation 12.2,

$$W_{t+1} = W_t - \bar{m}_t \left(\frac{\eta}{\sqrt{V_t} + \in} \right) \tag{12.2}$$

where ε is the small value to avoid divide by zero error when $v_t = 0$

Every output unit summates its weighted input signal and the corresponding output signal is computed using the activation function as shown in Equation 12.3:

$$O(t) = \left(\Sigma h(t) W_0 + B_0 \right) \tag{12.3}$$

Every neuron in the network has an activation function. The activation function used for hidden units is Rectified Linear Unit (RELU) and the activation function used in the output layer is Softmax. RELU outputs zero for inputs less than zero and output one for inputs greater than zero. The Softmax activation function is applied to the output of the hidden layers to convert the scores into probability values that sum up to one.

The softmax activation function can be expressed as shown in Equation 12.4:

$$\sigma(\vec{Z}) = \frac{e^{Z_i}}{\sum_{j=1}^{K} e^{Z_j}} \tag{12.4}$$

12.2.3.5 Performance metrics

The major objective of this section is to evaluate the machine learning model for achieving the greatest accuracy during the testing process. The performance measures that are evaluated for classifier efficiency and performance using the outcomes of the model such as True Positive (TP), False Positive (FP), True Negative (TN), and False Negative (FN). These measures can be calculated from the counts obtained from the results of testing samples belonging to class 'C'. By identifying the total number of correct results made among all the predictions, classification accuracy scores can be calculated as shown in Equation 12.5:

$$\text{Accuracy} = \frac{TP + TN}{TP + TN + FP + FN} \tag{12.5}$$

The ratio of predicted positive results to actual positive events is known as precision.

$$Precision = \frac{\sum_{i=0}^{n} TP}{\sum_{i=1}^{n} TP + FP} \qquad (12.6)$$

During the result phase, the following error measures are determined.

- The average of the squared differences between the expected predictions and actual prediction is represented by the Mean Square Error (MSE) as shown in Equation 12.7.

$$MSE = \frac{1}{N} \sum_{i-1}^{N} (S_{oi} - S_{ei})^2 \qquad (12.7)$$

Where S_o stands for the actual value and S_e stands for the expected value

- The main purpose of the machine learning model is to predict the results and to understand how close the predicted values are to the actual values. RMSE indicates the average difference between the predicted values model and the expected values as shown in Equation 12.8.

$$RMSE = \sqrt{\frac{1}{n} \sum (S_{oi} - S_{ei})^2} \qquad (12.8)$$

12.2.3.6 Data storage and indexing

Spark converts the streaming data into data frames and the data is stored in HBase. The results of the ML model, and the original data, are then saved in HBase. The recommendation output of the ML model is indexed with the Elasticsearch cluster.

The work steps can be summarized as follows:

Step 1   Users search for specified content through the User interface;
Step 2   If there is hit information in Elasticsearch, go to Step 7, else go to Step 3;
Step 3   The service issues the input through the message subscription module in Kafka;
Step 4   The Spark stream processing module receives the input;
Step 5   The Spark stream processing module processes it with the deployed machine learning module;
Step 6   The Spark Stream processing module stores the input and output in the HBase storage;

Step 7   The Spark Stream processing module stores the processed data indexed with the field survey number in Elasticsearch;

Step 8   Return the crop recommendation to the User interface.

12.3 EXPERIMENTAL RESULTS

The suggested work is run on an Ubuntu 18.04 LTS platform with a Core i5 CPU and 8GB RAM. JDK 11, Apache Kafka 2.12, and Apache Spark 2.1.0 are utilized to create streaming processing in the system. *pyspark. streaming* and *pyspark. MLLIB* is the package to build the offline machine learning model and apply the deployed model on real-time data. HBase version 2.2.1 is used for data storage. First, the soil, and weather data of a particular location is collected and the price data for the crops are collected from different data sources cited in Table 12.1. The data retrieved is in the form of CSV files. Kafka was used for reading the data that was produced from soil characteristics, weather parameters, and pricing details of the crops. Kafka uses a python producer script to stream the data saved in the CSV file as data streams into the analytics model in Spark. When this script is run, data from the CSV file is read and published into Kafka at regular intervals. This information is published under appropriate Kafka topics. The data ingested is used by Spark streaming to build the machine learning model using the MLlib package.

The dataset available is in the form of categorical data. The categorical data is converted into numerical form using one-hot-encoding that enables the easy implementation of machine learning algorithms. In the course of developing the machine learning model, the complete data set is deliberately split in the ratio of 70:30 as the training set and the testing set. The Artificial Neural Network model consists of 17 input neurons which represent the different attributes in the data set. As the number of hidden layers to be used cannot be determined, the model has to be trained with the different number of hidden layers. The output layer consists of 4 neurons to represent the suitability classes of rice, maize, finger millet, and sugarcane. Neural networks are trained with three, five, and seven hidden layers. Each neural network is trained for 50 epochs.

The model is trained with a stochastic gradient descent algorithm and is evaluated with different optimization techniques, namely the momentum method, RMSProp, and Adam Optimizer [30]. The performance of the model is shown in Table 12.3. The Rectified Linear Unit (RELU) activation function was employed in the input layer and hidden layers. This model evaluates the probability of crops, namely maize, finger millets, rice, and sugarcane, and ranks them in order of suitability. Therefore the softmax function was used in the output layer.

The suggested model classifies the experimental data set into four classes: highly suitable (class 1), moderately suitable (class 2), marginally suitable

Table 12.3 Performance of artificial neural networks with different optimization algorithms

	Accuracy
Momentum method	87.2
RMS prop	89.1
Adam optimizer	91.3

(class 3), and not suitable (class 4). Class 1 and 2 agriculture sites might be used for crop cultivation in their current state; however, class 3 agriculture farms need to be further processed by applying sufficient manure before cultivating the crop, and class 4 agriculture land cannot be used for cultivation.

One of the most important architectural parameters that influence the execution of ANN during the training and preparation of information for learning is the number of hidden layers and the number of epochs. In this regard, defining these parameters in advance has remained a challenge in machine learning research. In the proposed work, ANN was implemented with 3 and 5 hidden layers with 80 epochs respectively. Table 12.4 describes the results obtained for each class using ANN.

Table 12.5 defines the performance metrics of the decision tree classifier

From the results, we can infer that the performance of ANN with 5 hidden layers is better than the decision tree to forecast the suitability of the crops. Hence ANN can be used in the crop recommendation framework.

The crop suitability results are stored in the Elasticsearch cluster and are indexed to quickly access the results for the query posted through the web interface. The field survey numbers, along with the longitude and latitude values of the location, are considered as the index whose corresponding crop suitability is the document. Every index has the suitability for rice, maize, finger millet, and sugarcane.

Table 12.4 Performance metrics of ANN with 3 and 5 hidden layers iterated for 80 epoch

Performance metrics	Performance of ANN with three hidden layers				Performance of ANN with five hidden layers			
	Class 1	Class 2	Class 3	Class 4	Class 1	Class 2	Class 3	Class 4
Accuracy	0.86	0.87	0.81	0.84	0.88	0.83	0.91	0.81
Precision	0.87	0.85	0.88	0.95	0.91	0.94	0.89	0.95
MSE	0.012	0.01	0.01	0.001	0.001	0.014	0.001	0.001
RMSE	0.26	0.27	0.26	0.26	0.04	0.03	0.041	0.02

Table 12.5 Performance metrics of decision tree

| Performance metrics | Decision tree classifier | | | |
	Class 1	Class 2	Class 3	Class 4
Accuracy	0.78	0.72	0.73	0.84
Precision	0.77	0.74	0.78	0.75
MSE	0.014	0.02	0.024	0.002
RMSE	0.28	0.37	0.26	0.26

Table 12.6 Elasticsearch index information

Index	Document
Survey no 77, 13° 20'N, 77° 31' E	Suitability of Rice: Class 3 Suitability of Finger millet: Class 2 Suitability of Maize: Class 2 Suitability of sugarcane: Class 3
Survey no 78, 13° 20'N, 77° 31' E	Suitability of Rice: Class 3 Suitability of Finger millet: Class 1 Suitability of Maize: Class 2 Suitability of sugarcane: Class 2

The sample of the Elasticsearch cluster is shown in Table 12.6.

The suggested model analyses the given agriculture area and delivers reliable results for sustainable agriculture development. As a result, this proposed model might be employed as a land suitability recommendation model to boost crop production for long-term agriculture development.

12.4 DISCUSSION

The classification results obtained, as shown in Tables 12.4 and 12.5, are largely influenced by several performance parameters. For the multiclass classification data set, the suggested study aggregates several performance parameters for the evaluation and assessment of ANN and decision trees. The performance of ANN is found to increase with the increase in the number of hidden layers. The MSE and RMSE are found to be low and decreasing. Furthermore, this leads to a better convergence of error, which in turn lead to proper feature learning. The performance parameters discussed earlier, such as accuracy, precision and others, appear to follow a similar pattern.

The performance of ANN with five hidden layers is found to be significantly better than that of ANN with three layers. However, because the number of hidden layers has increased architectural complexity, convergence optimization can be accomplished with fewer iterations. On observing the performance measures of ANN with five hidden layers, it is found to be providing better results than the decision tree.

Furthermore, given the current experimental methodology, slightly different results could be obtained because the biases and initial weights of the neural networks are generated at random. If an appropriate weight set is fixed at the start, similar settings may yield better outcomes. This argument may be supported towards the end of the neural network training process and gradient descent may not always ensure a near-ideal weight set.

In comparison to other approaches, the suggested model analyses the given agriculture area and produces better results for sustainable agricultural production. This methodology helps agriculturists assess their land effectively by providing a reliable decision on the appropriateness level of agricultural land in four different categories. As a result, this proposed model might be employed as a land appropriateness recommendation model to boost crop production for long-term agriculture development.

12.5 CONCLUSION

In this chapter, we presented a real-time crop recommendation model that was developed based on Apache Kafka, Apache Spark, and Elasticsearch. The recommendation model reads the streaming data of soil characteristics for location-specific survey numbers, real-time weather data, and pricing details for the considered crops. The real-time soil and weather data is received by Apache Kafka and passed to the Apache Spark consumer. Artificial Neural Networks and decision tree learning algorithms were applied to the streaming soil, weather, and crop price data to derive meaningful insights for crop recommendation. With the measured accuracy values, it was learned that the recommendation model created with ANN performs better with an accuracy of 89 percent, compared to the decision tree classifier with 77 percent accuracy. The recommended suitable crops corresponding to input are indexed in Elasticsearch to enable the faster retrieval of results. The multiclass classification would provide real-time results to ensure better crop yields. The suggested system can be expanded in the future to account for market demand, market infrastructure availability, projected profit, post-harvest storage, and processing technologies. This would result in a detailed crop recommendation based on geographical, environmental, and economic factors, resulting in an effective agricultural system.

REFERENCES

[1] P. Slavin, "Climate and famines: A historical reassessment," *Wiley Interdiscip. Rev. Clim. Change*, vol. 7, no. 3, pp. 433–447, 2016, doi: 10.1002/wcc.395.

[2] P. C. Robert, "Precision agriculture: A challenge for crop nutrition management," *Plant Soil*, vol. 247, no. 1, pp. 143–149, 2002, doi: 10.1023/A:1021171514148.

[3] B. Basso, J. T. Ritchie, F. J. Pierce, R. P. Braga, and J. W. Jones, "Spatial validation of crop models for precision agriculture," *Agric. Syst.*, vol. 68, no. 2, pp. 97–112, 2001, doi: 10.1016/S0308521X(00)00063-9.

[4] V. Karimi, E. Karami, and M. Keshavarz, "Climate change and agriculture: Impacts and adaptive responses in Iran," *J. Integr. Agric.*, vol. 17, no. 1, pp. 1–15, 2018, doi: 10.1016/S2095-3119(17)61794-5.

[5] S. Wolfert, L. Ge, C. Verdouw, and M. J. Bogaardt, "Big Data in Smart Farming – A review," *Agric. Syst.*, vol. 153, pp. 69–80, 2017, doi: 10.1016/j.agsy.2017.01.023.

[6] N. Kshetri, "The emerging role of Big Data in key development issues: Opportunities, challenges, and concerns," *Big Data Soc.*, vol. 1, no. 2, 2014, doi: 10.1177/2053951714564227.

[7] R. Lokers, R. Knapen, S. Janssen, Y. van Randen, and J. Jansen, "Analysis of Big Data technologies for use in agro-environmental science," *Environ. Model. Softw.*, vol. 84, pp. 494–504, 2016, doi: 1016/j.envsoft.2016.07.017.

[8] A. M. S. Osman, "A novel big data analytics framework for smart cities," *Futur. Gener. Comput. Syst.*, vol. 91, pp. 620–633, 2019, doi: 10.1016/J.FUTURE.2018.06.046.

[9] "Open Government Data (OGD) Platform India." https://data.gov.in/ (accessed August 1, 2021).

[10] "World Bank Open Data | Data." https://data.worldbank.org/ (accessed August 1, 2021).

[11] "Home Page." https://www.krishimaratavahini.kar.nic.in/department.aspx (accessed August 1, 2021).

[12] "Bhoomi NBSS&LUP Geo Portal." https://www.nbsslup.in/bhoomi/ (accessed August 1, 2021).

[13] "IMD | Home." https://mausam.imd.gov.in/ (accessed July 6, 2021).

[14] "Arkansas Plant Disease database." https://www.uaex.edu/yard-garden/resource-library/diseases/ (accessed August 1, 2021).

[15] K. Adnan and R. Akbar, "An analytical study of information extraction from unstructured and multidimensional big data," *J. Big Data 2019 61*, vol. 6, no. 1, pp. 1–38, 2019, doi: 10.1186/S40537019-0254-8.

[16] P. Mikalef, I. O. Pappas, J. Krogstie, and M. Giannakos, "Big data analytics capabilities: a systematic literature review and research agenda," *Inf. Syst. E-bus. Manag. 2017 163*, vol. 16, no. 3, pp. 547–578, 2017, doi: 10.1007/S10257-017-0362-Y.

[17] S. Pudumalar, E. Ramanujam, R. H. Rajashree, C. Kavya, T. Kiruthika, and J. Nisha, "Crop recommendation system for precision agriculture," in *2016 8th International Conference on Advanced Computing, ICoAC 2016*, June 2017, pp. 32–36, doi: 10.1109/ICoAC.2017.7951740.

[18] R. Veerachamy and R. Ramar, "Agricultural Irrigation Recommendation and Alert (AIRA) system using optimization and machine learning in Hadoop for sustainable agriculture," *Environ. Sci. Pollut. Res. 2021*, pp. 1–20, March 2021, doi: 10.1007/S11356-021-13248-3.

[19] A. C. Fabregas, "Decision Tree Algorithm Applied in Suitability Assessment of Temporary Crops Based on Agrometeorological Forecasts," *Indian J. Sci. Technol.*, vol. 12, no. 26, pp. 1–7, 2019, doi: 10.17485/ijst/2019/v12i26/145096.

[20] "Apache Kafka." https://kafka.apache.org/ (accessed August 3, 2021).

[21] "Apache Spark," Accessed: July 30, 2021. [Online]. Available: https://run awayhorse001.github.io/LearningApacheSpark/setup.html

[22] X. Lu, D. Shankar, S. Gugnani, and D. K. D. K. Panda, "High-performance design of apache spark with RDMA and its benefits on various workloads," *Proc. – 2016 IEEE Int. Conf. Big Data, Big Data 2016*, pp. 253–262, 2016, doi: 10.1109/BIGDATA.2016.7840611.

[23] I. Triguero, M. Galar, D. Merino, J. Maillo, H. Bustince, and F. Herrera, "Evolutionary undersampling for extremely imbalanced big data classification under apache spark," *2016 IEEE Congr. Evol. Comput. CEC 2016*, pp. 640–647, November 2016, doi: 10.1109/CEC.2016.7743853.

[24] "Apache HBase – Apache HBase™ Home." https://hbase.apache.org/ (accessed August 6, 2021).

[25] V. A. Zamfir, M. Carabas, C. Carabas, and N. Tapus, "Systems monitoring and big data analysis using the elasticsearch system," *Proc. - 2019 22nd Int. Conf. Control Syst. Comput. Sci. CSCS 2019*, pp. 188–193, May 2019, doi: 10.1109/CSCS.2019.00039.

[26] P. E. Black, "Revisiting the Thornthwaite and Mather water balance," *J. Am. Water Resour. Assoc.*, vol. 43, no. 6, pp. 1604–1605, 2007, doi: 10.1111/j.1752-1688.2007.00132.x.

[27] L. G. K. Naidu, V. Ramamurthy, O. Challa, Rajendra Hegde, and P. Krishnan, "Soil suitability criteria for major crops," *National Bureau of Soil Survey and Land Usage Planning, Nagpur, India*, March 2006. http://krishikosh.egranth.ac.in/displaybitstream?handle=1/2034266 (accessed Jun, 30, 2020).

[28] J. Madhuri and M. Indiramma "Hybrid filter and wrapper methods based feature selection for crop recommendation," pp. 247–252, 2022, doi: 10.1109/ICESIC53714.2022.9783542.

[29] Z. Wang et al., "Spatiotemporal variability of reference evapotranspiration and contributing climatic factors in China during 1961–2013," *J. Hydrol.*, vol. 544, pp. 97–108, 2017, doi: 1016/j.jhydrol.2016.11.021.

[30] J. Zhang, "Gradient Descent based Optimization Algorithms for Deep Learning Models Training," 2019, [Online]. Available: http://arxiv.org/abs/1903.03614

[31] T. A. W. Mykel and J. Kochenderfer, "Algorithms for Optimization (The MIT Press) Free Pdf Download | Education Books," 2019, Accessed: July 24, 2022. [Online]. Available: https://edubookpdf.com/computer/algorithms-for-optimization-the-mit-press.html

Chapter 13

Application of artificial intelligence in small and medium enterprises

An overview

Laxmi Pandit Vishwakarma and Rajesh Kr Singh

Management Development Institute, Gurgaon, India

13.1 INTRODUCTION

Industry 4.0 technologies are now shaping business transformations, especially for small and medium-sized enterprises (SMEs), to the next level (Pelletier and Cloutier, 2019; Szedlak et al., 2020). Dwivedi et al. (2021) highlighted that SMEs have started adopting modern technologies such as AI. Forbes (2018) highlighted that 51% of SMEs feel that AI technology is essential. The AI in SMEs is expected to grow at a CAGR of 22.10% and reach $90.68 billion by 2027 (IndustryARC, 2022). AI technology has transformed the working landscape of SMEs (Hansen and Bøgh, 2021). AI technology is also regarded as critical for future economic growth (Grashof and Kopka, 2022). Adopting AI technology in SMEs improves their productivity and helps them to achieve a competitive advantage (Chan et al., 2018; Kumar and Kalse, 2021; Kundu et al., 2019). The application of AI helps SMEs to find new opportunities and make operational changes (Warner and Wäger, 2019). SMEs staff having higher technical skills and capabilities are responsible for the implementation of AI practices in the SMEs (Agostini and Nosella, 2016; Ghobakhloo and Ching, 2019). Drydakis (2022) revealed that AI technology empowered SMEs to enhance their capabilities. Pandya and Kumar (2022) highlighted that the adoption of Industry 4.0 technologies, such as AI, helps firms improve sustainability. AI technology allows SMEs to drive radical innovations (Grashof and Kopka, 2022). SMEs are a driving factor in removing poverty and producing high-quality outputs (Singh et al., 2010). Past studies highlight that SMEs lack the adoption of digital technologies and therefore need to upgrade themselves in the context of the adoption of Industry 4.0 technologies (Chatterjee et al., 2021; Pandey et al., 2020).

The definition of SMEs varies from country to country (Singh and Kumar, 2020). The European Commission has defined an SME as a company with less than 250 people, an annual balance sheet of less than €43 million, and a firm's total turnover of less than €50 million (European Commission, 2019). According to the World Bank, SMEs play an essential role in developing countries and are principally responsible for economic growth.

DOI: 10.1201/9781003462163-13

SMEs contribute to about 90% of total business and 50% of employment worldwide (World Bank 2021). Critical characteristics of SMEs from past literature revealed that SMEs are the key contributors to the country's employment generation and economic development. However, SMEs face problems handling fluctuating demands, managing a skilled workforce, and having resource and financial constraints (Grashof and Kopka, 2022; Putra and Santoso, 2020; Roffia et al., 2021; Singh and Kumar, 2020). McKinsey (2021) has defined SMEs based on three characteristics. Government can support SMEs based on these three characteristics. They are as follows:

Confidence: Government can help SMEs gain confidence by supporting them in the ease of business. The government can create a healthy and friendly environment, supporting entrepreneurship.

Growth: The government can support SMEs by providing them with financial incentives and technical assistance.

Competitiveness: Government can motivate SMEs to adopt digital technologies by providing them with digital infrastructural support, helping them to create a competitive environment.

The SMEs are primarily responsible for any economic growth. Roffia et al. (2021) highlight that small businesses create the highest number of jobs in the US and other parts of the world. Putra and Santoso (2020) examined that the nature of SME businesses consists of adaptive to change, are small in number, and may be receptive. Mogaji et al. (2020) observed that SMEs are increasing their working activities to create new business opportunities and expand themselves across different business markets. Implementing AI in SMEs encourages firms to adopt new ideas (Grashof and Kopka, 2022). Chakraborty and Biswas (2019) highlighted that SMEs play an essential role in the national and international markets. Similarly, Azevedo and Almeida (2021) examined the digital transition of SMEs in Europe. Wang et al. (2021) analyzed the intelligent transformation of SMEs in China. Most recently, Quansah et al. (2022) examined the adaptive practices in SMEs.

The study conducted a comparative case study of Canadian and American firms to understand the underlying dynamic capabilities of the firm. The identified dynamic capabilities of SMEs include communicating effectively, leveraging reciprocal relationships, process improvements, and continuous learning. Baabdullah et al. (2021) examined the application of AI in SMEs in Middle East countries. Azman et al. (2021) analyzed the benefits of adopting AI for automated bookkeeping for Malaysian SMEs. Ali Abbasi et al. (2022) propose a dual-stage analysis examining the growing role of social media marketing in SMEs. The study analyses the technological, organizational and environmental (TOE) factors for understanding the factors influencing social media marketing. Putra and Santoso (2020) also studied the TOE factors and the resource-based view theory to analyze the performance impact of e-business in Indonesian SMEs. Chatterjee et al. (2021) studied the digital transformation of SMEs in the Indian context. The study examined the moderating role of AI-based customer relationship management capability. Kundu et al. (2019) highlighted that SMEs make a positive contribution to the national economy. Maier (2016) revealed that SMEs face difficulties sustaining in the competitive and volatile market. These challenges are: access to lack of market information, resource constraints, limited workforce, fund shortage, limited availability of technologies, and lack of knowledge sharing (Lee et al., 2021). However, AI technology is responsible for mitigating these challenges and improving knowledge sharing or information flow (Alsheibani et al., 2020; Dwivedi et al., 2021 Ghobakhloo and Ching, 2019). Baabdullah et al. (2021) studied the antecedents and consequences of adopting AI in SMEs for B2B practices.

The endless benefits of AI have made SMEs adopt AI technology in multiple domains. AI technology helps business to improve their responsiveness and efficiency in the market. The adoption of AI technology effectively overcomes the challenges faced by SMEs. AI technology is responsible for leveraging the data, providing valuable insights from them, and thereby improving the firm performance (Ghobakhloo and Ching, 2019). With the usage of AI technology, SMEs' can easily manage their financial and non-financial performances (Baabdullah et al., 2021). Ayyagari et al. (2014) observed that implementing AI technology in SMEs makes interactions smooth among the organization's stakeholders. Agostini and Nosella (2016) highlighted that AI improves customer interactions and provides better customer services. Haenlein and Kaplan (2019) revealed that AI technology has the potential to collect data from different sources and analyses them. The comments on the data help in the decision-making processes, thereby benefiting the organization. Different applications of AI, such as dynamic pricing, automation, prediction, and optimization, help SMEs with multiple business functions. These activities were carried out manually in the past, but drawing on the use of AI technology, companies can directly run the algorithms and prepare future business plans. Further, it provides the flexibility to validate large data sets and alter situations using optimization methods. Alternate situations offer a broad understanding of the situations, helping the business partners to deliver high-quality targets, prepare plans and solve significant problems (Dwivedi et al., 2021; Haenlein and Kaplan, 2019). AI technology benefits SMEs by improving customer experiences, high revenue, high productivity, reliability, and agility (Alsheibani et al., 2020; Dwivedi et al., 2021; Rauch et al., 2019). Current literature lacks any discussion of AI applications in SMEs across various business domains in a single work. Therefore, our study examines SMEs' applications across business domains, such as finance, marketing, supply chain management, human resources, and business performance. Based on this, we aim to develop our research question:

RQ1: What are the applications of AI in SMEs across various business domains?

To answer our research question, the rest of the chapter is organized as follows. First, Section 13.2 of the chapter discusses the research methodology. Then, Section 13.3 examines the research findings and is divided into two sections. Subsection 13.3.1 highlights the descriptive analysis, and Subsection 13.3.2 illustrates the applications of AI in SMEs across various business domains. Finally, Section 13.4 highlights the study's conclusion, limitations, and future scope.

13.2 RESEARCH METHODOLOGY

The study follows the research methodology by Cruz-Cárdenas et al. (2021), which is familiar to the literature review. Initially, we started with the development of research questions. Then, based on the search query, databases and keywords were selected. The database selected for the study is the Scopus database. The keywords were searched in the string using AND and OR operators in the title, keyword, and abstract sections. Keywords such as "artificial intelligence" or "AI" and "SME" or "small and medium enterprises" were used. Initially, we found 390 documents on 8 November 2022. Further, using the inclusion and exclusion criteria, we selected the subject area as the decision sciences, business management, and social sciences. To have a holistic understanding of the topic, we decided on articles, conference proceedings, and book chapters. We further considered the works published in the "English" language. Having applied these criteria, we found 127 documents. Later, after reading the articles and abstracts, we described and summarized our findings. The findings of our study are discussed in the next section. Figure 13.1 shows the flow chart of the research methodology.

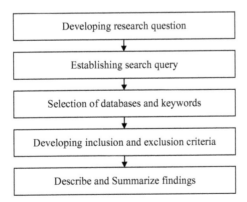

Figure 13.1 Flow chart of research methodology.

Source: Adapted from Cruz-Cárdenas et al., (2021).

13.3 RESEARCH FINDINGS

13.3.1 Descriptive analysis

A total of 127 documents were analyzed. The study highlights that there were relatively few publications on this topic at the start of this period. After 2017, however, there was a jump in the studies conducted on this topic. For example, in 2017, only one document was published, whereas, in 2018, 2019, and 2020, there were 3, 8, and 19 documents published, respectively. Further, in 2021, 30 papers were published; in 2022, up to 8 November 2022, 22 documents were published on this topic. The annual publications are shown in Figure 13.2.

The majority of the research documents selected for this study are conference papers (49%) and articles (46%), showing how recent the study was published. The rest, 5%, are book chapters. Figure 13.3 shows the document type considered for this study.

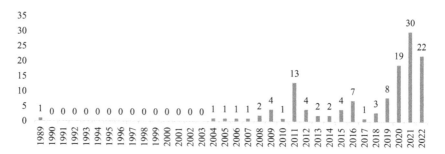

Figure 13.2 Annual Publications (*n* = 127).

Source: Author's contribution.

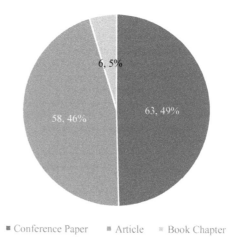

Figure 13.3 Document type (*n* = 127).

Source: Author's contribution.

13.3.2 Applications of AI technology in SMEs

Past studies reveal that the application of AI technology in SMEs has been observed across various business domains. These business domains are finance, marketing, supply chain, and human resource management. Therefore, this chapter also provides insights into the applications of AI across different business performances. The applications of AI technology in SMEs across other business functions is shown in Figure 13.4. The summarization of the applications of AI technology in SMEs is shown in Table 13.1.

13.3.2.1 Finance

AI technology provides various applications in the finance domain among SMEs. For example, it helps reduce fraudulent transactions, does financial planning, assess risks, evaluates credits, bookkeeping, and across different accounting activities—this section discusses the application of AI in the finance department of SMEs.

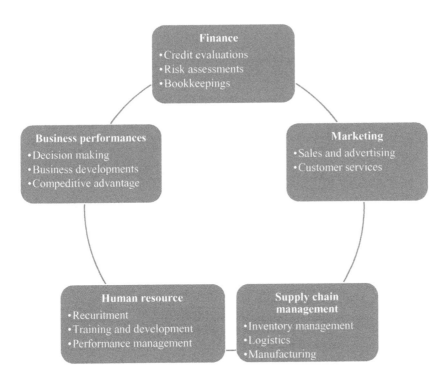

Figure 13.4 Applications of AI technology in SMEs.

Source: Author's contribution.

Table 13.1 Applications of AI technology in SMEs

Applications of AI in SMEs		
Finance		
Credit Evaluations	Analyze the creditworthiness of the borrowers	Kim et al. (2021)
	Analyses credit score and credit history	OECD (2021)
	Promote financial inclusion	OECD (2021)
	Contributes to portfolio management	OECD (2021)
	Automates reconciliations	OECD (2021)
Risk Assessments	Identify the anomalies in financial data	Dadteev et al. (2020); Yang et al. (2019)
	Suggests investment strategies	Dadteev et al. (2020); Yang et al. (2019)
	Predict insolvency	Camacho-Miñano et al. (2015)
	Monitor risk factors daily, weekly, and annually	OECD (2021)
	Unlock data insights, contributing to assessing risks	OECD (2021)
Bookkeeping	Automates the bookkeeping processes	Azman et al. (2021)
	Makes the process straightforward and on a real-time basis	Azman et al. (2021)
Marketing		
Sales and Advertising	Enhances product offerings	Drydakis (2022); Hansen and Bøgh (2021); Ulrich et al. (2021)
	Makes pricing strategies	Drydakis (2022); Hansen and Bøgh (2021); Ulrich et al. (2021)
	Pushes personalized notifications based on the customer's behaviour, online activities, past activities, and social media profiles	Drydakis (2022); Hansen and Bøgh (2021); Ulrich et al. (2021)
	Identify customer needs and divide them into different categories	Davenport et al. (2020); Fan et al. (2020)
	Helps in targeting potential customers by offering them multiple discounts and promo codes	Davidsson et al. (2018)
	Customizes the company's website based on customer preferences	Kumar et al. (2019)
	Calculates the advertising impacts	Davenport et al. (2020)
	Introduce recommendation services to boost sales	Lee et al. (2021)
Customer Services	Helps in engaging customers	Prentice et al. (2020)
	Answers online customer's queries and generates feedback forms	Prentice et al. (2020)
	AI-based chatbots enhance the communication between the firm and the customers	Ikumoro & Jawad (2019)
	Analyzes customer patterns	Ikumoro & Jawad (2019)
	Picks up the most crucial customer queries	Anshari et al. (2019)

(Continued)

Table 13.1 (Continued)

Applications of AI in SMEs

Supply chain management

Inventory Management	Helps demand forecasting	Vishwakarma et al., (2023); Borodavko et al. (2021)
	Reduce supply chain errors by 30-40%	Borodavko et al. (2021)
	Creates visibility and helps supply chain professionals to track inventory	Vishwakarma et al., (2023); Borodavko et al. (2021)
	Helps the firm build a secure IT network wherein the clients and the vendor organizations can coordinate smoothly	Bangalore Seetharam (2020)
Logistics	Suggests vehicle rerouting and delivering goods in shorter durations	Quansah et al. (2022)
	Helps in tracking the delivery of the product at each stage helps in tracking the delivery of the product at each stage	Vishwakarma et al., (2023); Vishwakarma and Singh, (2022)
	Helps reduce time, effort, and cost while delivering the product to the customer	Vishwakarma et al., (2023); Borodavko et al. (2021)
	Inspect products and processes	Vishwakarma and Singh, (2022)
	Automates the unloading process	Vishwakarma and Singh, (2022)
	Facilitates shipment tracking activities	Vishwakarma and Singh, (2022)
Manufacturing	Predicts the machine's maintenance time	Quansah et al. (2022); Vishwakarma and Singh, (2022)
	Helps in accurate forecasting	Quansah et al. (2022); Borodavko et al. (2021); Ulrich et al. (2021)
	Catches manufacturing defects	Quansah et al. (2022); Borodavko et al. (2021); Ulrich et al. (2021)
	Reduces unplanned downtime	Borodavko et al. (2021); Ulrich et al. (2021)
	Promotes agile production planning	Borodavko et al. (2021);
	Avoids production stoppage	Quansah et al. (2022); Ulrich et al. (2021)
	Manages demand shocks and transportation disruptions	Vishwakarma and Singh, (2022)
	Automates service orders from time to time	Quansah et al. (2022); Borodavko et al. (2021); Ulrich et al. (2021)
	Quality controls	Ghobakhloo and Ching (2019)

(Continued)

Table 13.1 (Continued)

Applications of AI in SMEs

Human Resources

Recruitment	Analyze the candidate's profile and scan each job application	Vishwakarma and Singh (2023); Johansson and Herranen (2019)
	Simplify the resume-scanning processes	Vishwakarma and Singh (2023); Johansson and Herranen (2019)
	Matches the right candidate for the job opening	Vishwakarma and Singh (2023); Johansson and Herranen (2019)
	Assess employee referrals	Vishwakarma and Singh (2023); Jia et al. (2018)
	Provides the employee details	Jia et al. (2018)
	Automates the onboarding processes	Vishwakarma and Singh (2023); Kaushal et al. (2021)
	Answers common queries during application filling	Jia et al. (2018)
	Smoothens the orientation processes	Vishwakarma and Singh (2023); Basu et al. (2022)
	Smoothen the employee internal mobility process	Basu et al. (2022)
Training and Development	Suggests the best training activities	Johansson and Herranen, (2019)
	Connect past employees' work experiences and examine their interest in future roles	Vishwakarma and Singh (2023); Kaushal et al. (2021)
Performance Management	Smoothens the entire performance measurement process	Kaushal et al. (2021);
	Provides results based on factual information	Vishwakarma and Singh (2023); Johansson and Herranen, (2019)
	Enable continuous assessments	Basu et al. (2022)
	Engage employees more effectively	Kaushal et al. (2021); Johansson and Herranen, (2019)

Business Performance

Decision-Making	Advances the data management process	Drydakis (2022); Dwivedi et al. (2021)
	Enhances the decision-making process	Vishwakarma and Singh (2023); Kim et al. (2021); Drydakis (2021)
	Attracts the decision-maker's interest	Dwivedi et al. (2021)
Business Developments	Helps the firm in running different simulations	Drydakis (2022); Baabdullah et al. (2021)
	Helps in reshaping the business developments	Basu et al. (2022); Baabdullah et al. (2021)
	Enhances organizational arrangements	Basu et al. (2022)
	Optimally analyses the organization's resources, market demand, and capability	Dwivedi et al. (2021); Baabdullah et al. (2021)

(Continued)

Table 13.1 (Continued)

Applications of AI in SMEs		
Competitive Advantage	Contributes to picking sales lead	Basu et al. (2022)
	Enhances customer service	Kumar and Kalse (2021); Kundu et al. (2019)
	Streamlines the business operations	Dwivedi et al. (2021); Baabdullah et al. (2021)
	Reduces the administrative burden	Johansson and Herranen (2019); Chan et al. (2018)
	Increases speed and consistency of various internal and external operations	Dwivedi et al. (2021); Baabdullah et al. (2021)
	Reduces customer waiting time	Baabdullah et al. (2021)
	Identify fraudulent activities	Drydakis (2022)
	Performs real-time optimization	Drydakis (2022); Dwivedi et al. (2021)

13.3.2.1.1 Credit evaluations

The AI algorithms deeply analyze the creditworthiness of the borrowers, improving the decision-making process (Kim et al., 2021). The algorithms classify the clients based on their credit scores and credit history (OECD, 2021) and, based on it, promote financial inclusion. It is also of further help in portfolio management. AI-based models help in credit allocation and, based on the customer's portfolio, either decline or accept the prospective borrowers. Neural network algorithms simplify the data analysis and make them understood by the clients, thereby promoting better communication between the client and the financial service provider. In addition, AI algorithms manage manual reconciliations and automate them (OECD, 2021).

13.3.2.1.2 Risk assessments

Neural network algorithms derived from AI technology help SMEs maximize cash flows and prevent liquidity risks. These algorithms allow the SMEs to identify the anomalies in financial data and, based on it, suggest the firm develop investment strategies (Dadteev et al., 2020; Yang et al., 2019). Camacho-Miñano et al. (2015) examined that AI algorithms help SMEs to predict insolvency in risk management. The dynamic nature of AI algorithms adjusts themselves based on the changing market or economic conditions (OECD, 2021). The AI and machine learning models monitor risk factors daily, weekly, and annually, thereby enhancing the client portfolio. Furthermore, AI algorithms manage structured and unstructured data and unlock insights from it quickly (OECD, 2021).

13.3.2.1.3 Bookkeeping

Bookkeeping is the recording of daily transactions from invoices until payments. It is a process of documenting each transaction to create a collection of accounting documents. In particular, it includes classifying each item (trades) correctly and entering it in real time. The manual record-keeping system needs special attention in managing all the paper-based receipts and journal entries. Azman et al. (2021) highlighted that AI technology automates the bookkeeping systems in SMEs. AI uses optical character recognition and pattern recognition to recognize documents. Machine learning further helps in auto journal record entries. It reduces errors, increases customer satisfaction, reduces claims rejections, works in real time, and improves the firm's revenue. After that, AI makes the bookkeeping processes more straightforward and more efficient at all levels of business.

13.3.2.2 Marketing

Rahman et al. (2021) and Basri (2020) highlighted that the adoption of AI in SMEs improves different areas of marketing. For example, AI technology improves social media marketing, sales, and advertisement, helps target customers, enables pricing strategies, and improves customer services. This section discusses the application of AI in the marketing management of SMEs.

13.3.2.2.1 Sales and advertising

The AI-based tools analyze the sales data (based on the locations (regions, states, countries)), festival seasons, and other factors. The analysis of historical sales data helps in enhancing product offerings and making pricing strategies. Based on historical data, AI technology further uses predictive analysis to analyze future scenarios, which allows SMEs to gain a competitive advantage. The AI-based algorithms give personalized notifications based on the customer's behaviour, online activities, past activities, and social media profiles (Drydakis, 2022; Hansen and Bøgh, 2021; Ulrich et al., 2021). Basri (2020) analyzed that adopting AI-based social media marketing improves SMEs' performances. Rahman et al. (2021) revealed that adopting AI enhances the firm's marketing capabilities AI-based algorithms, natural language processing, or social sentiment analysis help SMEs identify customer needs and divide the customers into different categories (Davenport et al., 2020; Fan et al., 2020). By classifying customers into different categories, SMEs can easily find and target potential customers through advertisements, offering them multiple discounts and promo codes (Davidsson et al., 2018). Davenport et al. (2020) examined that AI algorithms help SMEs to calculate the advertising impacts, thereby improving operational and purchase efficiency. Lee et al. (2021) highlight that many SMEs introduce recommendation services to boost sales. Studies reveal that if the right products or services are recommended to the customers, continuous purchases increases, and customers become loyal to the firm.

13.3.2.2.2 Customer services

AI technology helps in improving customer service. Prentice et al. (2020) examined that AI plays a crucial role in customer engagement. The AI tools help to answer online questions and generate feedback forms. Thus, allowing the firms to get customer feedback and to improve their performance. AI-based chatbots enhance communication between the firm and the customers. Ikumoro and Jawad (2019) highlighted that AI-based chatbots help respond to customer queries in an intelligent human conversation form. AI tools help in analyzing customer patterns. Therefore, it encourages more customized and personalized sales and marketing activities. AI chatbots help in identifying customer needs. It further communicates with potential customers and offers responses to common questions. AI-based algorithms pick up the most crucial customer queries. It recognizes the sentences such as "I'll leave your business" and "I would like to buy the product next week". AI voice bots help to contact thousands of customers at a single time and helps in responding the simple question. Hence, SMEs using chatbots can save time and effort (Anshari et al., 2019). Using AI-based chatbots, firms can offer world-class services, even with fewer human resources. Chatbots promote virtual communication, which both improves customer services and promotes sales (Pérez et al., 2019).

13.3.2.3 Supply chain management

Vishwakarma and Singh (2022) examined AI's applications among different supply chain stakeholders. Singh and Kumar (2020) conducted a study in the Indian SMEs context and revealed that effective supply chain management drives SMEs in a dynamic world economy. The applications of SMEs in supply chain management exists from suppliers, manufacturers, retailers, and customers. This section discusses the application of AI in the supply chain management of SMEs.

13.3.2.3.1 Inventory management

Managing inventory is an essential part of supply chain management. It helps to fulfill the customer's demand, reduces losses, and builds a positive customer relationship. AI helps demand forecasting, which further helps in managing inventories. The applications of AI technology help to reduce supply chain errors by 30–40%. AI algorithms include cycle times, temperatures, errors, quantities, lead times, and planned downtime. Adding such variables increases the accuracy. Such accuracy helps the firms to minimize lost sales. AI creates visibility and helps supply chain professionals to track inventory (Borodavko et al., 2021). By adopting AI technology, SMEs can free up the budgets and resources previously required for inventory control. Bangalore Seetharam (2020) highlighted that SMEs could improve their inventory management even during the COVID-19 pandemic using AI technology. AI technology helped the firm build a secure IT network wherein the clients and the vendor organizations can coordinate smoothly.

13.3.2.3.2 Logistics

AI collects internal and external data such as demographics and weather forecasts and provides insights (Quansah et al., 2022). AI algorithms suggest rerouting and delivering goods in shorter durations based on these insights. AI navigates different markets based on macro- and micro-economic factors. The technology helps in tracking the delivery of the product at each stage. In contrast, the product has left the manufacturing firm and is kept at warehouses during the in-transit stage and while the product is delivered (Vishwakarma and Singh, 2022). All the data are analyzed in real time; therefore, the company can take immediate necessary actions if any problem occurs. Such measures help better customer communication and provide faster delivery services. The algorithms used by AI also allow SMEs to reduce time, effort, and cost while delivering the product to the customer (Borodavko et al., 2021). AI technology helps in vehicle routing and scheduling, inspecting products and processes, automating the unloading process, and facilitating shipment tracking activities (Vishwakarma and Singh, 2022).

13.3.2.3.3 Manufacturing

AI technology uses predictive analysis, which helps predict the machine's maintenance time. Machine learning and expert systems help the firm to develop different algorithms, thereby helping in accurate forecasting. The forecasting techniques help the firm to analyze the historical patterns, the organization's resources, and customer demand and to manufacture the products based on these factors. AI helps to catch manufacturing defects and reduces the level of unplanned downtime. The applications of AI technology promote agile production planning. AI technology ensures that manufacturing is executed smoothly, avoiding production stoppage, demand shocks, and transportation disruptions (Vishwakarma and Singh, 2022). AI helps SMEs manage predictive maintenance during the manufacturing processes, automating service orders from time to time, monitoring quality controls, and identifying manufacturing defects (Borodavko et al., 2021; Ghobakhloo and Ching, 2019; Quansah et al., 2022; Ulrich et al., 2021).

13.3.2.4 Human resources

Jia et al. (2018) highlighted the six basic dimensions of human resource management and revealed that AI helps to manage all of them. These dimensions comprise: performance management, employee relationship management, training and development process, human resource strategy and planning, salary evaluation, and recruitment. In a similar vein, Kaushal et al. (2021) observed that AI technology helps organizations manage. These dimensions include talent acquisition, selection, performance analysis, training and learning, management and retention, recruitment, and onboarding. This section discusses the application of AI in the human resource management of SMEs.

13.3.2.4.1 Recruitment

During recruitment processes, SMEs receive frequently receive more than a hundred job applications. It becomes difficult to check each one of them manually. AI tools analyze the candidate's profile and are able to scan each job application. The natural language process tools simplify the resume scanning processes and match the right candidate to the right job opening. AI tools analyze the profile based on the historic work experiences and suggest the candidate's future role in the new organization (Johansson and Herranen, 2019). AI tools help assess employee referrals and provide the employee details, providing the maximum number of referrals. AI automates the onboarding processes and answers common queries during the process of application filling (Jia et al., 2018). The AI-based algorithms help the firm manage all the candidate's onboarding documents (Kaushal et al., 2021). AI smoothens the orientation processes and helps the human resource team to save a lot of time. The algorithms match the employee profile for any new vacancies. AI algorithms further smoothen the employee internal mobility process (Basu et al., 2022).

13.3.2.4.2 Training and development

Employees and new recruitments often need training and development. AI smoothens the process of onboarding and training activities for the entire team. Based on the employee's skill sets, AI algorithms suggest the best training activities for them and help them to navigate the process of completing them (Johansson and Herranen, 2019). In addition, deep learning and neural networks connect past employees' work experiences and examine their interest in future roles (Kaushal et al., 2021). Past studies highlight that SMEs faces resource and financial constraints (Grashof and Kopka, 2022; Putra and Santoso, 2020; Roffia et al., 2021; Singh and Kumar, 2020) and, therefore, AI algorithms help the firm to effectively manage their employee and provide them with training and development (Basu et al., 2022).

13.3.2.4.3 Performance management

In SMEs, the AI algorithms remove the bias and provide insights based on the data. The employee's performance or productivity is measured based on multiple data sources. The data points smooth the entire performance measurement process and provide results based on factual information. The feedback provided during the performance management process is free from biases and, therefore, neutral for each employee. Such an environment creates trust among the employee and helps build a healthy working environment. The AI-based performance management systems help in reducing human errors, enable continuous assessments and provide real-time analysis, engage employees more effectively, create data based on projections, and enhance training and development activities (Basu et al., 2022; Kaushal et al., 2021; Johansson and Herranen, 2019).

13.3.2.5 Business performances

Dwivedi et al. (2021) highlighted that AI has the potential to increase business performance. AI technology enhances organizational arrangements and attracts the decision-maker's interest (Basu et al., 2022). Furthermore, AI optimally analyses the organization's resources, market demand, and capability and helps the business improve its performance (Baabdullah et al., 2021). This section discusses the application of AI in the business performance of SMEs.

13.3.2.5.1 Decision-making

Drydakis (2022) revealed that AI technology advances the data management process. AI helps to make decisions much faster, and with greater accuracy. Dwivedi et al. (2021) highlighted that SMEs had adopted AI technology to enhance the decision-making processes. AI drives conclusions, which helps the decision-makers to take decisions more rapidly (Drydakis, 2021; Kim et al., 2021). The AI algorithms provide insights from data at each stage. The algorithms further enable decision-makers to time slice the data at different levels, going from planning to execution. Therefore, it helps in making effective decision-making processes. It further accelerates the decision-making processes, having the firm save time, cost, and effort (Dwivedi et al., 2021).

13.3.2.5.2 Business developments

AI helps SMEs forecast and prioritize different activities, keeping control of various constraints. AI algorithms clean the data, structures them, and provide additional insights. By combing with other technologies, such as digital twin or blockchain, these insights can be made secure and different simulations can be run out of them. Such simulations help SMEs keep up the business developments, analyzing them and altering them whenever required. With the applications of AI, SMEs can make a strong comeback and reshape business developments (Baabdullah et al., 2021; Basu et al., 2022; Drydakis, 2022).

13.3.2.5.3 Competitive advantage

The AI technology automates multiple processes, makes the process less complicated, streamlines the business operations, reduces the administrative burden, improve the return on investments, enables high-efficiency marketing, provides real-time information, helps in picking sales leads, enhances customer service, increases the speed and consistency of various internal and external operations, reduces customer waiting time, identifies fraudulent activities, performs real-time optimization and therefore provides a competitive advantage to the SMEs adopting AI technology (Basu et al., 2022; Chan et al., 2018; Drydakis, 2022; Dwivedi et al., 2021; Johansson and Herranen, 2019; Kumar and Kalse, 2021; Kundu et al., 2019).

13.4 CONCLUSION, LIMITATIONS, AND FUTURE SCOPE

Organizations are now changing rapidly with the emergence of Industry 4.0 technologies, one of the most significant of which is Artificial Intelligence (AI). The adoption of AI in SMEs has been observed over recent years, but past literature lacks any discussion of the applications in a single piece of work. Our study aims to fill this gap; therefore, in this chapter we have provided the applications of AI in SMEs across different business functions. These functions include finance, marketing, supply chain management, human resources, and business performance. Under finance, we discussed the application of AI for credit evaluations, risk assessments, and bookkeeping. Under marketing, we discussed the application of AI for sales, advertising and customer services. Under supply chain management, we discussed the application of AI for inventory management, logistics, and manufacturing. Under human resources, we discussed the application of AI for recruitment, training and development, and performance management. Finally, under business performance, we discussed the application of AI for decision-making, business development, and competitive advantage. Our study will help future researchers to achieve an understanding of the various applications of AI in SMEs. Furthermore, the study will help AI practitioners develop business growth strategies. In addition, future studies can use other databases such as Web of Science or EBSCO host databases to conduct similar studies in this domain.

REFERENCES

Agostini, L., & Nosella, A. (2016). The central role of a company's technological reputation in enhancing customer performance in the B2B context of SMEs. *Journal of Engineering and Technology Management*, 42, 1–14.

Ali Abbasi, G., Abdul Rahim, N. F., Wu, H., Iranmanesh, M., & Keong, B. N. C. (2022). Determinants of SME's social media marketing adoption: Competitive industry as a moderator. *Sage Open*, 12(1). https://doi.org/10.1177/21582440211067220

Alsheibani, S., Cheung, Y., & Messom, C. (2020). Re-thinking the competitive landscape of artificial intelligence. In *Proceedings of the 53rd Hawaii international conference on system sciences* (pp. 5861–5870). HICSS, Hawaii, USA, January 7–10.

Anshari, M., Almunawar, M. N., Lim, S. A., & Al-Mudimigh, A. (2019). Customer relationship management and big data enabled: Personalization and customization of services. *Applied Computing and Informatics*, 15(2), 94–101.

Ayyagari, M., Demirguc-Kunt, A., & Maksimovic, V. (2014). Who creates jobs in developing countries? *Small Business Economics*, 43(1), 75–99.

Azevedo, A., & Almeida, A. H. (2021). Grasp the challenge of digital transition in SMEs—A training course geared towards decision-makers. *Education Sciences*, 11(4), 151.

Azman, N. A., Mohamed, A., & Jamil, A. M. (2021). Artificial intelligence in automated bookkeeping: A value-added function for small and medium enterprises. *JOIV: International Journal on Informatics Visualization*, 5(3), 224–230.

Baabdullah, A. M., Alalwan, A. A., Slade, E. L., Raman, R., & Khatatneh, K. F. (2021). SMEs and artificial intelligence (AI): Antecedents and consequences of AI-based B2B practices. *Industrial Marketing Management*, 98, 255–270.

Bangalore Seetharam, S. (2020). Developing a digital AI roadmap for retail. Master's thesis. Metropolia University of Applied Sciences.

Basri, W. (2020). Examining the impact of artificial intelligence (AI)-assisted social media marketing on the performance of small and medium enterprises: Toward effective business management in the Saudi Arabian context. *International Journal of Computational Intelligence Systems*, 13(1), 142–152.

Basu, S., Majumdar, B., Mukherjee, K., Munjal, S., & Palaksha, C. (2022). Artificial intelligence–HRM interactions and outcomes: A systematic review and causal configurational explanation. *Human Resource Management Review*, 33(1), 100893.

Borodavko, B., Illés, B., & Bányai, Á. (2021). Role of artificial intelligence in supply chain. *Academic Journal of Manufacturing Engineering*, 19(1), 75–79.

Camacho-Miñano, M. D. M., Segovia-Vargas, M. J., & Pascual-Ezama, D. (2015). Which characteristics predict the survival of insolvent firms? An SME reorganization prediction model. *Journal of Small Business Management*, 53(2), 340–354.

Chakraborty, D., & Biswas, W. (2019).,Evaluating the impact of human resource planning programs in addressing the strategic goal of the firm: An organizational perspective. *Journal of Advances in Management Research*, 16(5), 659–682.

Chan, L., Morgan, I., Simon, H., Alshabanat, F., Ober, D., Gentry, J., Min, D., & Cao, R. (2018). Survey of AI in cybersecurity for information technology management. In *IEEE technology and engineering management conference*. Atlanta: TEMSCON.

Chatterjee, S., Chaudhuri, R., Vrontis, D., & Basile, G. (2021). Digital transformation and entrepreneurship process in SMEs of India: A moderating role of adoption of AI-CRM capability and strategic planning. *Journal of Strategy and Management*, 15(3), 416–433.

Cruz-Cárdenas, J., Zabelina, E., Guadalupe-Lanas, J., Palacio-Fierro, A., & Ramos-Galarza, C. (2021). COVID-19, consumer behavior, technology, and society: A literature review and bibliometric analysis. *Technological Forecasting and Social Change*, 173, 121179.

Dadteev, K., Shchukin, B., & Nemeshaev, S. (2020). Using artifcial intelligence Technologies to predict cash flow. *Procedia Computer Science*, 169, 264–268.

Davenport, T. H., Guha, A., Grewal, D., & Bressgott, T. (2020). How artificial intelligence will change the future of marketing. *Journal of the Academy of Marketing Science*, 48(1), 24–42.

Davidsson, P., Recker, J., & von Briel, F. (2018). External enablement of new venture creation: A framework. *Academy of Management Perspectives*, 34(3), 311–332.

Drydakis, N. (2021). Mobile applications aiming to facilitate immigrants' societal integration and overall level of integration, health and mental health does artifcial intelligence enhance outcomes? *Computers in Human Behavior*, 117(April), 106661.

Drydakis, N. (2022). Artificial Intelligence and reduced SMEs' business risks. A dynamic capabilities analysis during the COVID-19 pandemic. *Information Systems Frontiers*, 24(4), 1223–1247.

Dwivedi, Y. K., Hughes, L., Ismagilova, E., Aarts, G., et al. (2021). Artificial intelligence (AI): Multidisciplinary perspectives on emerging challenges, opportunities, and agenda for research, practice and policy. *International Journal of Information Management*, 57, 101994.

European Commission. (2019). *User guide to the SME definition. Internal market, industry, entrepreneurship and SMEs, European Commission, Brussels*. Brussels, Belgium: European Commission.

Fan, X., Ning, N., & Deng, N. (2020). The impact of the quality of intelligent experience on smart retail engagement. *Marketing Intelligence and Planning*, 38(7), 877–891.

Forbes. (2018). 10 Stats you don't know about small businesses. *Salesforce essential*. Available at: https://www.forbes.com/sites/salesforceessentials/2018/11/01/10-stats-you-dont-know-about-small-businesses/?sh=6faa0d6c608d. Last accessed: 11th September 2022.

Ghobakhloo, M., & Ching, N. T. (2019). Adoption of digital technologies of smart manufacturing in SMEs. *Journal of Industrial Information Integration*, 16, 100107.

Grashof, N., & Kopka, A. (2022). Artificial intelligence and radical innovation: An opportunity for all companies?. *Small Business Economics*, 1–27.

Haenlein, M., & Kaplan, A. (2019). A brief history of artificial intelligence: On the past, present, and future of artificial intelligence. *California Management Review*, 61(4), 5–14.

Hansen, E. B., & Bøgh, S. (2021). Artifcial intelligence and Internet of Things in small and medium-sized enterprises: A survey. *Journal of Manufacturing Systems*, 58(B), 362–372.

Ikumoro, A. O., & Jawad, M. S. (2019). Assessing intelligence conversation agent trends-chatbots-AI technology application for personalized marketing. *TEST Engineering and Management*, 81, 4779–4785.

IndustryARC. 2022. Artificial intelligence in small and medium business market overview. Available at: https://www.industryarc.com/Report/17911/artificial-intelligence-market-in-small-medium-business.html. Last accessed: 13th November, 2022.

Jia, Q., Guo, Y., Li, R., Li, Y., & Chen, Y. (2018). A conceptual artificial intelligence application framework in human resource management. In *Proceedings of The 18th International Conference on Electronic Business* (pp. 106–114). ICEB, Guilin, China, December 2–6.

Johansson, J., & Herranen, S. (2019). The application of artificial intelligence (AI) in human resource management: current state of AI and its impact on the traditional recruitment process [bachelor thesis].: Jönköping University; 2019 May. URL: http://www.diva-portal.org/smash/get/diva2:1322478/FULLTEXT01.pdf [last accessed 14-7-2023]

Kaushal, N., Kaurav, R. P. S., Sivathanu, B., & Kaushik, N. (2021). Artificial intelligence and HRM: Identifying future research agenda using systematic literature review and bibliometric analysis. *Management Review Quarterly*, 73, 455–493.

Kim, M. S., Lee, C. H., Choi, J. H., Jang, Y. J., Lee, J. H., Lee, J., & Sung, T. E. (2021). A study on intelligent technology valuation system: Introduction of KIBO patent appraisal system II. *Sustainability*, 13(22), 12666.

Kumar, A., & Kalse, A. (2021). Usage and adoption of artifcial intelligence in SMEs. *Materials Today: Proceedings*. https://doi.org/10.1016/j.matpr.2021.01.595

Kundu, S. C., Mor, A., Kumar, S., & Bansal, J. (2019)., Diversity within management levels and organizational performance: Employees' perspective. *Journal of Advances in Management Research*, 17(1), 110–130.

Lee, K. J., Hwangbo, Y., Jeong, B., Yoo, J., & Park, K. Y. (2021). Extrapolative collaborative filtering recommendation system with Word2Vec for purchased product for SMEs. *Sustainability*, 13(13), 7156.

Maier, E. (2016). Supply and demand on crowdlending platforms: Connecting small and medium-sized enterprise borrowers and consumer investors. *Journal of Retailing and Consumer Services*, 33, 143–153.

McKinsey. (2021). Unlocking growth in small and medium-size enterprises. Available at: https://www.mckinsey.com/industries/public-and-social-sector/our-insights/unlocking-growth-in-small-and-medium-size-enterprises. Last accessed: 8th November 2022.

Mogaji, E., Soetan, T. O., & Kieu, T. A. (2020). The implications of artificial intelligence on the digital marketing of financial services to vulnerable customers. *Australasian Marketing Journal*. https://doi.org/10.1016/j.ausmj.2020.05.003

OECD. (2021). Artificial intelligence, machine learning and big data in finance: Opportunities, challenges, and implications for policy akers. Available at: https://www.oecd.org/finance/artificial-intelligence-machine-learningbig-data-in-finance.htm. Last accessed: 8th November 2022.

Pandey, N., Tripathi, A., Jain, D., & Roy, S. (2020), Does price tolerance depend upon the type of product in e-retailing? Role of customer satisfaction, trust, loyalty, and perceived value. *Journal of Strategic Marketing*, 28(6), 522–541.

Pandya, D., & Kumar, G. (2022). Applying Industry 4.0 technologies for the sustainability of small service enterprises. *Service Business*, 17(1), 37–59.

Pelletier, C., & Cloutier, L. M. (2019, January). Challenges of digital transformation in SMEs: Exploration of IT-related perceptions in a service ecosystem. In *Proceedings of the 52nd Hawaii international conference on system sciences*, Band 52 (pp. 4967–4976).

Pérez, P., De-La-Cruz, F., Guerrón, X., Conrado, G., Quiroz-Palma, P., & Molina, W. (2019). ChatPy: Conversational agent for SMEs. In *2019 14th Iberian conference on information systems and technologies (CISTI)* (pp. 1–6). Coimbra: CISTI. https://doi.org/10.23919/CISTI.2019.8760624

Prentice, C., Weaven, S., & Wong, I. A. (2020). Linking AI quality performance and customer engagement: The moderating effect of AI preference. *International Journal of Hospitality Management*, 90, 102629.

Putra, P. O. H., & Santoso, H. B. (2020). Contextual factors and performance impact of e-business use in Indonesian small and medium enterprises (SMEs). *Heliyon*, 6(3), e03568.

Quansah, E., Hartz, D. E., & Salipante, P. (2022). Adaptive practices in SMEs: Leveraging dynamic capabilities for strategic adaptation. *Journal of Small Business and Enterprise Development*, 29(7), 1130–1148.

Rahman, M. S., Hossain, M. A., & Fattah, F. A. M. A. (2021). Does marketing analytics capability boost firms' competitive marketing performance in data-rich business environment?. *Journal of Enterprise Information Management*, 35(2), 455–480.

Rauch, E., Dallasega, P., & Unterhofer, M. (2019). Requirements and barriers for introducing smart manufacturing in small and medium-sized enterprises. *IEEE Engineering Management Review*, 47(3), 87–94.

Roffia, P., Moracchiato, S., Liguori, E., & Kraus, S. (2021). Operationally defining family SMEs: A critical review. *Journal of Small Business and Enterprise Development*, 28(2), 229–260.

Singh, R. K., Garg, S. K., &Deshmukh, S. G. (2010). The competitiveness of SMEs in a globalized economy: Observations from China and India. *Management Research Review*, 33(1), 54–65.

Singh, R. K., & Kumar, R. (2020). Strategic issues in supply chain management of Indian SMEs due to globalization: An empirical study. *Benchmarking: An International Journal*, 27(3), 913–932.

Szedlak, C., Poetters, P., & Leyendecker, B. (2020). Application of artificial intelligence in small and medium-sized enterprises. In *Proceedings at 5th NA international conference on industrial engineering and operations management*, Detroit, Michigan, USA, August 10–14, 2020.

Ulrich, P. Frank, V., & Kratt, M. (2021). Adoption of artifcial intelligence technologies in German SMEs – Results from an empirical study. In S. Hundal, A. Kostyuk, & D. Govorun (Eds.), *Corporate governance: A search for emerging trends in the pandemic times* (Vol. 189, pp. 76–84). Dubai: PACIS. https://doi.org/10.22495/cgsetpt13

Vishwakarma, L. P., & Singh, R. K. (2022). Application of artificial intelligence (AI) in supply chain: An overview. *Artificial intelligence of things for smart green energy management* (pp. 191–212). Part of the Studies in Systems, Decision and Control book series (SSDC), Volume 446.

Vishwakarma, L. P., & Singh, R. K. (2023), An analysis of the challenges to human resource in implementing artificial intelligence. In Tyagi, P., Chilamkurti, N., Grima, S., Sood, K., & Balusamy, B. (Eds.), *The adoption and effect of artificial intelligence on human resources management, part B (Emerald Studies in Finance, Insurance, and Risk Management)*, Emerald Publishing Limited, Bingley, pp. 81–109.

Vishwakarma, L. P., Singh, R. K., Mishra, R., & Kumari, A. (2023). Application of artificial intelligence for resilient and sustainable healthcare system: Systematic literature review and future research directions. *International Journal of Production Research*, 1–23. SI-Artificial Intelligence Applications in Healthcare Supply Chain Networks under Disaster Condition.

Wang, J., Lu, Y., Fan, S., Hu, P., & Wang, B. (2021). How to survive in the age of artificial intelligence? Exploring the intelligent transformations of SMEs in central China. *International Journal of Emerging Markets*, 17(4), 1143–1162.

Wang, W., Deng, S. T., Tan, C.-W., & Pan, Y. (2019). Smart generation system of personalized advertising copy and its application to advertising practice and research. *Journal of Advertising*, 48(4), 356–365.

Warner, K. S. R., & Wäger, M. (2019). Building dynamic capabilities for digital transformation: An ongoing process of strategic renewal. *Long Range Planning*, 52, 326–349.

World Bank. (2021). Small and medium enterprises (SME) finance. Available at: https://www.worldbank.org/en/topic/smefinance. Last accessed: 8th November 2022.

Yang, X., Mao, S., Gao, H., Duan, Y., & Zou, Q. (2019). Novel financial capital flow forecast framework using time series theory and deep learning: A case study analysis of Yu'e Bao transaction data. *IEEE Access*, 7, 70662–70672.

Chapter 14

Industry 4.0 and sustainable supply chain

A review and future research agenda

Ashok Yadav, Rajiv Kumar Garg, Anish Sachdeva and Sheetal Soda

National Institute of Technology, Jalandhar, India

14.1 INTRODUCTION

The term "Industry 4.0," often known as the "fourth industrial revolution," defines a prospective model for the manufacturing sector (Birkel & Müller, 2021). The German government introduced Industry 4.0 (I.4.0) in 2011 to promote factory automation. Many industries have rethought their structures in consideration of the shifting wants of consumers, the uncertainty of supply and demand, the importance of environmental concerns, and the scarcity of economic rewards available to businesses in certain areas (Naseem & Yang, 2021). As a result of government regulations and consumer demand, economic, environmental, and social factors must all be considered in the manufacturing processes.

The aim of I4.0 is to promote sustainability by facilitating the use of automation across all systems involved in the manufacturing process (Jamwal et al., 2021b). Industries are currently searching for new technology adoption to meet sustainability standards, and several alternatives, including "Big Data Analytics", "Blockchain technology", "Cyber-Physical Systems", "Cloud Computing", and "Circular Economy", have all proved popular (Kumar Mangla & Luthra, 2018). Businesses in today's interconnected world must adopt sustainable development strategies that benefit society, the economy, and the environment. To gain an edge in the market, businesses work toward becoming more environmentally friendly. Therefore, digital manufacturing systems encourage companies to prioritize sustainability and innovation, two essential drivers of the global economy (Fatorachian & Kazemi, 2021). This is why businesses of all sizes in the manufacturing sector are working on strategies and regulations to make their production processes greener. Unfortunately, the current state of knowledge and experience is insufficient to integrate sustainability fully into the SCM (Naz et al., 2022). I4.0 is now a global movement to increase the effectiveness of SCM and the effectiveness of operations, which calls for a shift in how goods are created and used. I4.0 would be a turning point in achieving the sustainability goal, even if it were not widely discussed before adopting the 2030 Agenda (Fatorachian & Kazemi, 2021). It can be a

DOI: 10.1201/9781003462163-14

meeting place for those interested in bringing the Sustainable Development Goals (SDGs) into line with the ongoing digital transformation. To date, few studies have been conducted that describe the role of I4.0 technologies in achieving supply chain sustainability and the effect these technologies have on the supply chains of manufacturing organizations (Jamwal et al., 2021a). Innovation firms have been experimenting with digitalizing industrial organizations throughout the previous ten years. Due to the technologies' significant influence on the industry, research into them has also begun. Integrating these technologies into the supply chain system is a critical concern for industrial businesses. But much study has already been done to describe Industry 4.0, its underlying principles, and its connection to manufacturing firms (Luthra & Mangla, 2018).

The Scopus database is being considered as a resource in this research. I4.0 and the sustainable supply chain (SSC) were used to find articles included in the study-mapping process from 2017 through 2023. This chapter explores the potential benefits of an SSC in the fourth industrial revolution. The following questions are posed for further exploration in the present investigation:

RQ1 What are the citation structure and publication trends in this field of study?

RQ2 How can I4.0 technologies fit into a sustainable supply chain?

RQ3 Which Universities or authors are collaborating in the fields of Industry 4.0 and sustainable supply chains?

There were four main parts of the chapter. The first section discusses the study's introduction and research questions. In Section 14.2, we'll talk about the sustainable supply chain and its relevance to Industry 4.0. The bibliometric examination of the SSC in I4.0 is discussed in Section 14.3 of the article. Finally, the study's implications and conclusions are presented in Section 14.4.

14.1.1 Industry 4.0 and the sustainable supply chain

From a managerial standpoint, adopting new value chain innovations in technology and procedure is being actively pursued. When green, lean, dispersed manufacturing is combined with the latest IT-based Industry 4.0 efforts, sustainable culture is created in the industrial supply chain (Zhang et al., 2016). This will spur the emergence of innovative, environmentally friendly corporate practices, particularly in the developing world's industrial sector. Supply chain operations, corporate processes, and even business models may all be profoundly impacted by Industry 4.0 (Naseem & Yang, 2021; Tuffnell et al., 2019).

Sustainability and enhanced supply chain adaptability are two factors that must be incorporated into today's industrial systems (Kunkel et al., 2022; Patidar et al., 2022). For improved data transmission and control,

Industry 4.0 permits the creation of a worldwide cyberphysical network of machines, equipment, sensors, and facilities. A highly adaptable and intelligent worldwide cyberphysical network that facilitates a smart manufacturing and value chain. Improvements in product design, material and machine needs, product lifecycle and SCM, and so on have a multiplier effect on the company's total performance. In the supply chain framework, Industry 4.0 allows highly organized links between materials, goods, and equipment to meet client requirements (Hombach et al., 2018; Yazdani et al., 2020). With Industry 4.0 technology, crucial production metrics, including current production rate, energy usage, material flow, client orders, and supplier information, can be tracked and managed in real time (Balaman et al., 2016; de Camargo Fiorini et al., 2021). To boost efficiency and foster the growth of a more sustainable culture, industrial supply chains have recently begun to adopt the concepts of Industry 4.0 and sustainability (Agrawal et al., 2021; Dhamija & Bag, 2020). There is some hope that adopting a new "Industry 4.0" way of thinking will be helpful in advancing sustainable development in both the business world and the wider society.

Supply chains are often cited as another example of how Industry 4.0 will transform the world in the future (Hombach et al., 2018). The optimal use of resources, technology, etc., are just a few examples of how smart manufacturing can improve sustainability. For this reason, the current body of work is necessary to fully comprehend the sustainability orientation promoted by Industry 4.0 in supply chains. Businesses will benefit more from Industry 4.0 if their management has a firm grasp of what it means. Therefore, Industry 4.0 calls for a well-structured nomenclature and targeted study to establish a clear description (Singh et al., 2022; Teerasoponpong & Sugunnasil, 2022). Currently, many obstacles and opportunities are involved in creating a sustainable corporate environment within the context of Industry 4.0. A sustainable supply chain (SSC) prioritizes reducing or eliminating supply chain-related negative impacts through eco-efficient practices such as recycling, composting, and cutting-edge technologies (Singh et al., 2022). Many researchers have proposed definitions of SSCs over the years. One of the most common is this one, which states that: A sustainable supply chain reduces the adverse effects of the supply chain on the environment and maximizes the efficiency with which its resources are used (Gebhardt et al., 2021; Sulkowski, 2019).

Many factors contribute to the expansion of businesses, including the necessity to adopt new technologies and the incessant demands of consumers quickly. Increasing competition in today's markets has prompted companies to upgrade their production infrastructure to meet rising consumer demands for more specialized goods and services. Industry 4.0, which combines production and business procedures and brings together retailers and wholesalers, is one promising option (Frederico, 2021). IoT and cyberphysical systems can be used to solve the technical problems which plague Industry 4.0. Putting together CPS modules makes up "Industry 4.0" (Bal & Pawlicka, 2021). Industry 4.0 encompasses a broader view of the Industrial

Internet by incorporating fields as diverse as mining, healthcare, power generation, and manufacturing.

Increased global competitiveness is currently seen as a result of the adoption of digitalization and sustainability concepts in industries (Yadav et al., 2022). Improved efficiency, less waste, decreased energy consumption, optimized processes, reduced overproduction, new jobs, preventative maintenance, and the creation of jobs for people with disabilities are all among the outcomes of a sustainable Industry 4.0 (Gundu et al., 2022; Yadav et al., 2021).

14.1.2 Challenges to Industry 4.0 technology in sustainable supply chains

Table 14.1 Challenges while adopting Industry 4.0 in a sustainable supply chain

A failure to integrate several forms of technology	Integration of technology is crucial for efficient communication and increased output. The manufacturing sector is struggling to develop a versatile interface that can connect a wide range of disparate parts. The effective interchange and analysis of data in a manufacturing setting depends on the integration and support of many distinct components of cyberphysical networks. Therefore, it is crucial to create a system to incorporate technology for creating a powerful I4.0-driven SSC.	(Liao et al., 2022; Nayal et al., 2022)
Issues with working together and coordinating	In order to implement I4.0 concepts and improve supply chain sustainability, corporate stakeholders must work closely together and communicate openly. Coordination and collaboration with suppliers are crucial for better communication mechanisms due to the high compatibility challenges of hardware and software, which should necessitate standardized interfaces and synchronization of data to get better synchronization with manufacturers.	(Brintrup et al., 2022; Iftikhar et al., 2022)
Concerns about safety	Connectivity between different types of organizations is an aspect of Industry 4.0 that has the potential to streamline the production and distribution process. Unfortunately, however, attackers frequently take advantage of the supply chain systems' innate security flaws. The supplier is the root of many security issues, as they are susceptible to phishing and the theft of privileged credentials, both of which can lead to widespread data leakage. Supply chain chokepoints can compromise an entire organization since their effects are seen throughout the chain's dependent processes. For a manufacturing facility to become a smarter factor and a supply chain to become a smarter value chain, security must be prioritized.	(Aliahmadi et al., 2022; Ganesh & Kalpana, 2022; Liao et al., 2022; Saha et al., 2022)

(Continued)

Table 14.1 (Continued)

Insufficient knowledge of the effects of Industry 4.0	The implications of Industry 4.0 are poorly understood by the scientific community and business executives alike. Concentrated and well-organized research is necessary to provide a thorough definition of I4.0 in the literature. Managers in industry and practise recognise Industry 4.0's significance in the manufacturing sector, but they lack clarity on the direct impact it will have on their capacity to meet sustainability targets via existing supply networks.	(Aliahmadi et al., 2022; Bourke, 2019; Nimmy et al., 2022)
Problems with Industry 4.0 adoption due to insufficient research and development	The term "Industry 4.0" has been inferred in various ways by different types of working managers. Companies often lack the necessary precise decision strategies to effectively deploy Industry 4.0. A primary cause of this is the paucity of studies that have specifically addressed the many challenges associated with implementing I4.0.	(Nayal et al., 2022; Wang, 2022)

14.2 METHODOLOGY

For publications about Industry 4.0 and supply chains, a search of peer-reviewed journal and conference articles was put into place in September 2022. Following are the keywords that were used in the article search for the study area: (TITLE-ABS-KEY ("Industry 4.0") AND TITLE-ABS-KEY ("sustainable supply chain" OR "sustainable supply chain management" OR "Digital supply chain" AND "bibliometric review")).

The following is the study's list of inclusion criteria:

Publications written and published in English;
Articles dated before September 2023;
Publications must be either refereed journals or conference proceedings;
Articles centred on SSC and I4.0 technology;
All articles must be brief or comprehensive (not an editorial or abstract).

14.3 RESULTS

14.3.1 Basic information about data

Figure 14.1 compiles the most crucial data from scholarly articles about Industry 4.0 and sustainable supply chains. Data for papers chosen from Scopus-indexed journals between 2017 and 2023 are displayed. Figure 14.1 also shows the authors, co-authors, and document statistics of the total number of scientific papers published in the chosen journals. Figure 14.1 shows that the number of documents related to Industry 4.0 in the sustainable

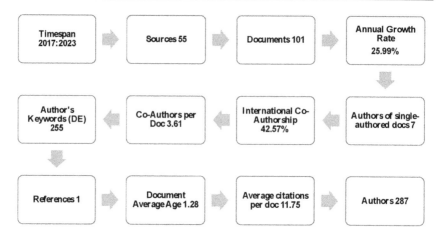

Figure 14.1 Basic information amount data.

supply chain has increased by 25.99% per year from a total of 101 papers written by 287 authors using 255 keywords.

14.3.2 Year-wise publication

Figure 14.2 shows how the number of articles published in these journals fluctuated between 2017 and 2023. Figure 14.2 shows that the number of papers published annually on I4.0 technology and SSC climbed gradually from 2017 to 2020 and then increased dramatically to more than 20

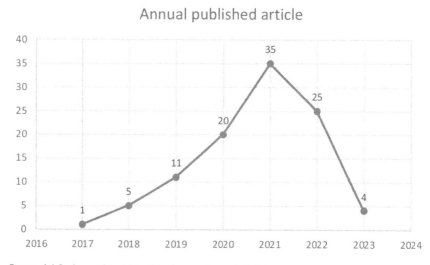

Figure 14.2 Annual publication in a selected field.

publications. This points to a growing interest in studies examining how I4.0 and SSC might be implemented in the real world.

14.3.3 Mean citation structure per year

The total number of citations for each year is presented in Table 14.2. As can be observed, the average number of citations received in a given year reached its maximum point in 2018, which was the year that saw the beginning of the integration of the idea of Industry 4.0 with a sustainable supply chain.

14.3.4 Top 10 authors impact

Listed in Table 14.3 is the author-provided citation format for previously published articles on the topic of I4.0 and SSC. The Scopus database shows

Table 14.2 Yearly mean citation

Year	N	MeanTCperArt	MeanTCperYear	CitableYears
2017	1	0.00	0.00	5
2018	5	106.00	26.50	4
2019	11	74.09	24.70	3
2020	20	41.20	20.60	2
2021	35	14.97	14.97	1
2022	25	1.48		0
2023	4	0.00	0.00	−1

Table 14.3 Top 10 authors' citation structure

Element	h_index	g_index	m_index	TC	NP	PY_start
BAG S	3	3	0.6	148	3	2018
JAYAKRISHNA K	3	3	0.75	392	3	2019
KUMAR S	3	3	1.5	79	3	2021
LUTHRA S	3	4	0.6	533	4	2018
MANAVALAN E	3	3	0.75	392	3	2019
MANGLA SK	3	5	0.6	537	5	2018
AGUAYO-GONZÁLEZ F	2	2	0.5	39	2	2019
ANTUCHEVICIENE J	2	2	1	26	2	2021
BELHADI A	2	2	1	65	2	2021
CHIAPPETTA JABBOUR CJ	2	3	0.667	83	3	2020

that MANGLA SK has the highest h-index and g-index of any of the top 10 authors, with 3 and 5, respectively, for 2018.

14.3.5 Top 10 affiliations with the publication

Figure 14.3 displays the top ten academic institutions contributing the most to the study of I4.0 in the context of SSC. Figure 14.3 shows that the Centre for Research and Technology University has the most publications in the Scopus database, with 11 total articles. After that comes Loughborough University, with eight articles.

14.3.6 Top 10 most frequent word

Figure 14.4 displays the results of a keyword analysis conducted to determine the most popular terms for studying I4.0 and SSC. SCM and sustainable development were two of the most frequently used words in the query, appearing 45 and 40 times, respectively. SSC and I4.0 are the other most popular search terms related to this concept. You can see the most popular search terms used to find publications relevant to a specific topic in Figure 14.4.

14.3.7 Most-cited countries

According to citation statistics in Figure 14.5, India, the United States, and France obtained the most citations; these are the top three most-cited countries in this research subject.

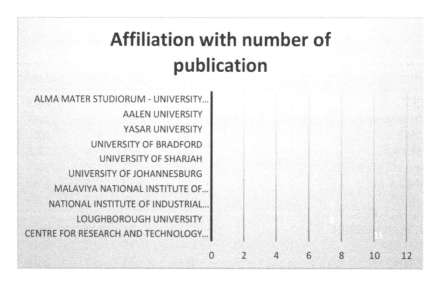

Figure 14.3 Top 10 organization with publication in this field.

Top 10 most frequent word

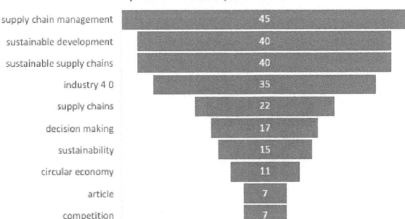

Figure 14.4 Top 10 keywords used in I4.0 and SSC.

Most Cited Countries

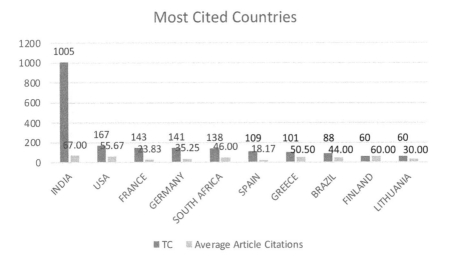

Figure 14.5 Top 10 cited countries in the given research field.

14.3.8 Word cloud

Figure 14.6 displays a word cloud including the most frequently used terms throughout the search for scholarly articles. When searching the database, use these terms to find publications that share commonalities in their abstract, title, and keywords. The magnitude of the word represents how often it appears in the chosen scientific papers (Figure 14.6). Industry 4.0,

WordCloud   ▶ Run

Plot   Table

Figure 14.6 Word cloud of keywords.

supply chains, sustainability, closed-loop supply chain, logistics, etc., are all centred around the term map as the essential concepts explored in this research.

14.3.9 Country's production over time

Figure 14.7 depicts the top 5 countries working in I4.0 and sustainable supply chain initiatives. Brazil has the most documents, then India, Morocco, Spain, and Turkey. Further research shows that only India, a developing country, ranks among the top five countries actively pursuing I4.0 and an SSC.

14.3.10 Top 10 most relevant journals

According to Figure 14.8, the *Journal of Cleaner Production* and *Sustainability* rank first in the number of scholarly papers published on the chosen research subject.

14.4 CONCLUSION AND FUTURE IMPLICATIONS

This chapter presents studies indicative of current research on integrating I4.0 technology into the supply chain process to achieve sustainability. This study summarizes the findings of 101 other studies conducted on Industry 4.0

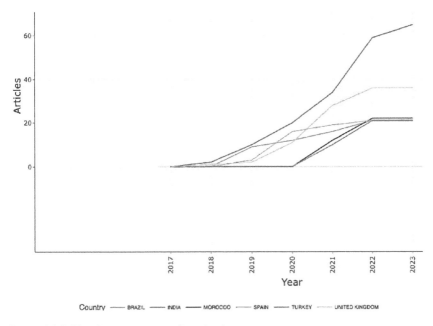

Figure 14.7 Top 6 countries working in the given area.

adoption in manufacturing organizations. Its conclusions are both intriguing and valuable for future researchers in the field. Many scholars have created various conceptual models and frameworks to better understand I4.0
and SSC. The results of this study revealed that there aren't many research
works that have combined supply chain sustainability with I4.0 technology. The *Journal of Cleaner Production* and *Sustainability* rank first and
second in the number of scholarly papers published in the chosen research
area. The result of the study also shows the annual growth rate of publication, which is 25.99%. The study also indicates that Centre for Research
and Technology University has the highest number of publications in the
Scopus database, with a total of 11 articles. After that comes Loughborough
University, with eight publications. The use of I4.0-based technologies has
revolutionized SSC operations. The significance of a comprehensive evaluation and review of studies on the subject of I4.0 and SSC has been highlighted in this study. We compiled a thorough literature evaluation in the
chosen area using bibliometric analysis. This research will inform managers
of the benefits of incorporating Industry 4.0 into a supply chain's efforts to
be more sustainable. More comprehensive reviews in future research could
include more aspects of the supply chain. The scope of this research is con-
fined to scholarly journals; future investigations could expand to incorporate additional types of papers. The review paper's focus on I4.0 and an SSC
raises several questions which are worthy of exploration.

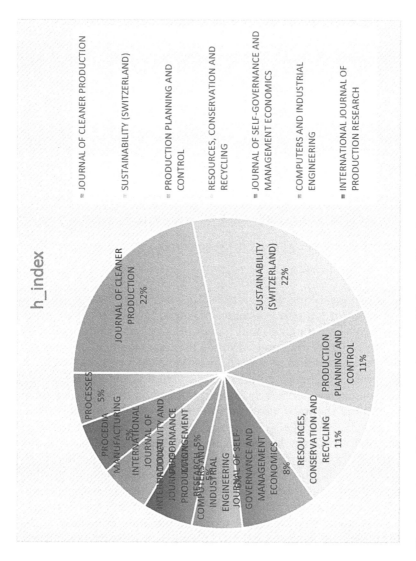

Figure 14.8 Top 10 relevant journals in the research area of Industry 4.0 and sustainable supply chain.

REFERENCES

Agrawal, R., Wankhede, V. A., Kumar, A., & Luthra, S. (2021). A systematic and network-based analysis of data-driven quality management in supply chains and proposed future research directions. *The TQM Journal*, 35, 73–101.

Aliahmadi, A., Nozari, H., & Ghahremani-Nahr, J. (2022). AIoT-based sustainable smart supply chain framework. *International Journal of Innovation in Management, Economics and Social Sciences*, 2(2), 28–38.

Bal, M., & Pawlicka, K. (2021). Supply chain finance and challenges of modern supply chains. *LogForum*, 17(1), 71–82.

Balaman, Ş. Y., Matopoulos, A., Wright, D. G., & Scott, J. (2016). Integrated optimization of sustainable supply chains and transportation networks for multi technology bio-based production: A decision support system based on fuzzy ε-constraint method. *Journal of Cleaner Production*, 172, 2594–2617. Scopus. https://doi. org/10.1016/j.jclepro.2017.11.150

Birkel, H., & Müller, J. M. (2021). Potentials of industry 4.0 for supply chain management within the triple bottom line of sustainability – A systematic literature review. *Journal of Cleaner Production*, 289, 125612.

Bourke, E. (2019). Smart production systems in industry 4.0: Sustainable supply chain management, cognitive decision-making algorithms, and dynamic manufacturing processes. *Journal of Self-Governance and Management Economics*, 7(2), 25–30. Scopus. https://doi.org/10.22381/JSME7220194

Brintrup, A., Kosasih, E. E., MacCarthy, B. L., & Demirel, G. (2022). Digital supply chain surveillance: Concepts, challenges, and frameworks. In *The Digital Supply Chain* (pp. 379–396). Elsevier.

de Camargo Fiorini, P., Chiappetta Jabbour, C. J., Lopes de Sousa Jabbour, A. B., & Ramsden, G. (2021). The human side of humanitarian supply chains: A research agenda and systematization framework. *Annals of Operations Research*, 1–26.

Dhamija, P., & Bag, S. (2020). Role of artificial intelligence in operations environment: A review and bibliometric analysis. *TQM Journal*, 32(4), 869–896. Scopus. https://doi.org/10.1108/TQM-10-2019-0243

Fatorachian, H., & Kazemi, H. (2021). Impact of Industry 4.0 on supply chain performance. *Production Planning & Control*, 32(1), 63–81.

Frederico, G. F. (2021). Towards a supply chain 4.0 on the post-COVID-19 pandemic: A conceptual and strategic discussion for more resilient supply chains. *Rajagiri Management Journal*.

Ganesh, A. D., & Kalpana, P. (2022). Future of artificial intelligence and its influence on supply chain risk management – A systematic review. *Computers & Industrial Engineering*, 108206.

Gebhardt, M., Kopyto, M., Birkel, H., & Hartmann, E. (2021). Industry 4.0 technologies as enablers of collaboration in circular supply chains: A systematic literature review. *International Journal of Production Research*, 1–29.

Gundu, K., Jamwal, A., Yadav, A., Agrawal, R., Jain, J. K., & Kumar, S. (2022). Circular economy and sustainable manufacturing: A bibliometric based review. In R. Agrawal, J. K. Jain, V. S. Yadav, V. K. Manupati, & L. Varela (Eds.), *Recent Advances in Industrial Production* (pp. 137–147). Springer. https://doi. org/10.1007/978-981-16-5281-3_13

Hombach, L. E., Büsing, C., & Walther, G. (2018). Robust and sustainable supply chains under market uncertainties and different risk attitudes – A case study of the German biodiesel market. *European Journal of Operational Research*, *269*(1), 302–312. Scopus. https://doi.org/10.1016/j.ejor.2017.07.015

Iftikhar, A., Ali, I., Arslan, A., & Tarba, S. (2022). Digital innovation, data analytics, and supply chain resiliency: A bibliometric-based systematic literature review. *Annals of Operations Research*, 1–24.

Jamwal, A., Agrawal, R., Sharma, M., Kumar, V., & Kumar, S. (2021a). Developing A sustainability framework for Industry 4.0. *Procedia CIRP*, *98*, 430–435.

Jamwal, A., Agrawal, R., Sharma, M., Manupati, V. K., & Patidar, A. (2021b). Industry 4.0 and sustainable manufacturing: A bibliometric based review. In *Recent Advances in Smart Manufacturing and Materials* (pp. 1–11). Springer.

Kumar Mangla, S., & Luthra, S. (2018). Evaluating challenges to Industry 4.0 initiatives for supply chain sustainability in emerging economies.

Kunkel, S., Matthess, M., Xue, B., & Beier, G. (2022). Industry 4.0 in sustainable supply chain collaboration: Insights from an interview study with international buying firms and Chinese suppliers in the electronics industry. *Resources, Conservation and Recycling*, *182*. Scopus. https://doi.org/10.1016/j.resconrec.2022.106274

Liao, M., Lan, K., & Yao, Y. (2022). Sustainability implications of artificial intelligence in the chemical industry: A conceptual framework. *Journal of Industrial Ecology*, *26*(1), 164–182.

Luthra, S., & Mangla, S. K. (2018). Evaluating challenges to Industry 4.0 initiatives for supply chain sustainability in emerging economies. *Process Safety and Environmental Protection*, *117*, 168–179.

Naseem, M. H., & Yang, J. (2021). Role of Industry 4.0 in supply chains sustainability: A systematic literature review. *Sustainability*, *13*(17), 9544.

Nayal, K., Kumar, S., Raut, R. D., Queiroz, M. M., Priyadarshinee, P., & Narkhede, B. E. (2022). Supply chain firm performance in circular economy and digital era to achieve sustainable development goals. *Business Strategy and the Environment*, *31*(3), 1058–1073.

Naz, F., Agrawal, R., Kumar, A., Gunasekaran, A., Majumdar, A., & Luthra, S. (2022). Reviewing the applications of artificial intelligence in sustainable supply chains: Exploring research propositions for future directions. *Business Strategy and the Environment*.

Nimmy, S. F., Hussain, O. K., Chakrabortty, R. K., Hussain, F. K., & Saberi, M. (2022). Explainability in supply chain operational risk management: A systematic literature review. *Knowledge-Based Systems*, *235*, 107587.

Patidar, A., Sharma, M., Agrawal, R., Sangwan, K. S., Jamwal, A., & Gonçalves, M. (2022). Sustainable supply chain research and key enabling technologies: A systematic literature review and future research implications. In J. Machado, F. Soares, J. Trojanowska, & V. Ivanov (Eds.), *Lecture Notes in Mechanical Engineering* (p. 319). Springer Science and Business Media Deutschland GmbH; Scopus. https://doi.org/10.1007/978-3-030-78170-5_27

Saha, E., Rathore, P., Parida, R., & Rana, N. P. (2022). The interplay of emerging technologies in pharmaceutical supply chain performance: An empirical investigation for the rise of Pharma 4.0. *Technological Forecasting and Social Change*, *181*, 121768.

Singh, D., Sharma, A., & Rana, P. S. (2022). Role of Industry 4.0 practices in supply chain resilience. *ECS Transactions*, *107*(1), 6607.

Sulkowski, A. (2019). Industry 4.0 era technology (AI, big data, blockchain, DAO): Why the law needs new memes. *Kan. JL & Pub. Pol'y Online*, *29*, 1.

Teerasoponpong, S., & Sugunnasil, P. (2022). Review on artificial intelligence applications in manufacturing industrial supply chain – Industry 4.0's perspective. In *2022 Joint International Conference on Digital Arts, Media and Technology with ECTI Northern Section Conference on Electrical, Electronics, Computer and Telecommunications Engineering (ECTI DAMT & NCON)* (pp. 406–411).

Tuffnell, C., Kral, P., Durana, P., & Krulicky, T. (2019). Industry 4.0-based manufacturing systems: Smart production, sustainable supply chain networks, and real-time process monitoring. *Journal of Self-Governance and Management Economics*, *7*(2), 7–12. Scopus. https://doi.org/10.22381/JSME7220191

Wang, H. (2022). Linking AI supply chain strength to sustainable development and innovation: A country-level analysis. *Expert Systems*, e12973.

Yadav, A., Jamwal, A., Agrawal, R., Manupati, V. K., & Machado, J. (2022). Environmental impact assessment during additive manufacturing production: Opportunities for sustainability and Industry 4.0. In *Smart and Sustainable Manufacturing Systems for Industry 4.0*. CRC Press.

Yadav, A., Yadav, A., Agrawal, R., & Jamwal, A. (2021). *Life Cycle Assessment and Life Cycle Energy Analysis of Buildings: A Review* (SSRN Scholarly Paper No. 3847316). https://papers.ssrn.com/abstract=3847316

Yazdani, M., Wang, Z. X., & Chan, F. T. S. (2020). A decision support model based on the combined structure of DEMATEL, QFD and fuzzy values. *Soft Computing*, *24*(16), 12449–12468. Scopus. https://doi.org/10.1007/s00500-020-04685-2

Zhang, S., Lee, C. K. M., Wu, K., & Choy, K. L. (2016). Multi-objective optimization for sustainable supply chain network design considering multiple distribution channels. *Expert Systems with Applications*, *65*, 87–99. Scopus. https://doi.org/10.1016/j.eswa.2016.08.037

Chapter 15

A comprehensive study of artificial intelligence in supply chains

Dinesh Kumar

Protiviti Middle East Member Firm, Dubai, United Arab Emirates

Nittin and Mahesh Chand

J C Bose University of Science & Technology, YMCA, Faridabad, India

15.1 INTRODUCTION

Artificial intelligence (AI) refers to the abilities of communication of machines to imitate human capabilities. Since AI technology is an emerging approach, various companies and manufacturing organizations are shifting their role from remote monitoring to control, autonomous AI-based systems and optimization systems in order to enhance their functionality [1, 5, 28]. AI also forms a more holistic perspective for industries' applicability in supply chain management (SCM) since the level of interest from researchers and practitioners is high, there is an immediate need to examine the AI contribution in the SCM domain [30, 31, 40]. Various studies have specified this need as of great importance. To overcome this gap, this study provides a literature review as well as by answering main research questions i.e.: how can AI contribute potentially to SCM studies? By considering the above theoretical definition, four key words are extracted primarily namely as, production, marketing, SCM, and logistics. The second section of this study defines the methodology of the research. The third section defines analysis and synthesis of research study. The fourth section provides the conclusions of this study. The fifth section provides the limitations and implications of this research study. The final section provides the future research work, including Sub-Research Questions (SRQ) propositions.

15.2 METHODOLOGY

This study is based on an evidence and systematic literature review (SLR) approach. To accomplish this, a five-step process is followed, as depicted in Figure 15.1 is used. This is further defined in the following sections:

DOI: 10.1201/9781003462163-15

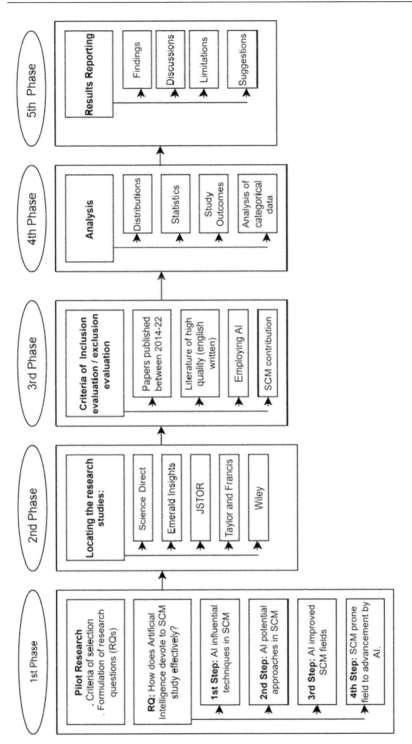

Figure 15.1 SLR research process.

15.2.1 Pilot research

a. *Pilot search*: This search is about the evaluations in the SCM field and the available literature (particularly from the electronic databases available, as mentioned in Table 15.1).

b. *The research question*: Since the formulation of the most prominent research questions acts as the main difficulty and the crucial part of the research design, it should be designed strategically. For instance, typical research questions might be worded as follows: How does AI potentially contribute to the SCM domain? To answer this, we further divided it into the main four sub-research questions (SRQs): Firstly, in this study, the most prominent AI approaches applicable in SCM studies are considered in order to obtain a better insight into the topic. In the context of answering the second SRQ, those prominent AI approaches which can be implemented in SCM research are analyzed. Third, SRQ utilized the tasks and subfields of AI in SCM. Finally, in order to answer the final SRQ, the tasks and subfields of AI in SCM which are to be improved by the applications of AI are analyzed for further research studies in the domain. The main point of the first and third SRQs is to examine the available literature and contribute deeper knowledge for both practitioners and researchers. By contrast,

Table 15.1 SLR search protocol

Databases	Articles searched	Searched fields	String searched	Timespan
Sciencedirect	Keywords, abstract and title	All	Key words "artificial intelligence in SCM" AND	2014–2022
Emerald Insights	Abstract, title and keywords	All	"Supply chain Management" and "AI in SCM"	2014–2022
JSTOR	Caption, abstract and title	Public, Business and Administration, Advertising and Marketing.	"AI role in Business, Marketing and Advertising"	2014–2022
Wiley library	Title, keywords and abstract	All	"Supply chain management"	2014–2022
Taylor and Francis	Keywords and Title	All	"Importance of AI in SCM"	2014–2022

the main aim of the second and fourth SRQs is to analyze possible opportunities and possible gaps for research and practical enhancement to formulate guidelines for future studies.

c. *Research sources*: The study in this chapter has drawn on five databases, i.e., Sciencedirect, Emerald Insights, JSTOR, Wiley Library, and Taylor & Francis, as these databases provide widespread coverage of peer-reviewed literature in the context of our main research questions, which are further explored using search strings as mentioned in Table 15.1.

d. *Criteria of Inclusive Evaluation/Exclusive Evaluation*: As specified in Figure 15.1, through the use of exclusion and inclusion criteria from our pilot search, we identified a total of 175 articles within the selected timespan, i.e., between 2014 and 2022, which contributed to discussions of potential trends and applications related to the role of AI in SCM. Second criterion intends to address quality and relevance: i.e., only conference papers and peer-reviewed journals were taken into consideration for the review, while chapters, discussion, news articles, and case studies are eliminated for consideration of this systematic literature review; in addition, each individual paper was thoroughly read by both authors in order to ensure the quality of the paper. The following selection criteria were applied: (1) Paper written in English only, (2) Applicability of a prominent AI tool in SCM (3) AI primarily contributing to SCM domain. Research papers selection was checked in accordance with the systematic literature review criteria of reputed journal papers and results were discussed and compared. Through the application of the selection criteria, the number of papers selected for synthesis and analysis was reduced to 52.

e. *Analysis*: In the analysis of these 52 articles, various specific characteristics in context to our research questions were formulated as follows: Study domain of the SCM field (i.e., production, logistics, marketing, and supply chain); prominent subfield(s) of study; AI approach(es) used; findings and outcomes. In the context of synthesis, we aim to define and describe the various relationships of the characteristics.

f. *Result Reporting*: In order to address an audience of academics and scholars, the results of this study were defined in term of statistics, discussions, and tabulations.

15.3 ANALYSIS AND SYNTHESIS

The next step after the collection of domain-related papers is data synthesis and data analysis. In the case of data analysis, the prime objective is to partition individual study into its constituent parts as well as describing the overall contacts and connections. Data synthesis works with the aim of identification of association between various studies. Both of the mentioned activities in content of current study are represented in the sections below.

15.3.1 Statistics and distribution

Out of the 52 articles considered for review, 10 papers were in the area of marketing, 11 papers concern logistics, 14 papers discuss production and the remaining 17 papers discuss the general field of supply chains. As shown in Figure 15.2, reviews covered the timespan 2014–2022. Literature was primarily being sourced as follows: 25% from conference proceedings and 75% from peer-reviewed journal papers (Figure 15.3).

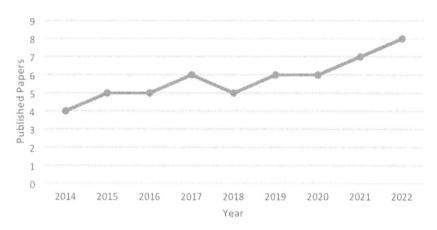

Figure 15.2 Time distribution.

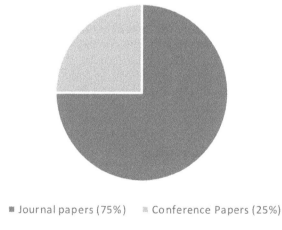

■ Journal papers (75%)   ▨ Conference Papers (25%)

Figure 15.3 Paper type distribution.

15.3.2 Categorical description of SLR

Table 15.2 selects the various papers in the context of SCM fields of marketing [22, 37, 3, 23], supply chain [3, 19, 26, 36, 46, 47, 50], production [24, 16] and logistics [4, 17, 20, 27, 33]. The next step is to content and subfield summarization. As mentioned in Table 15.2, a total of 10 articles are assigned to the marketing field whereas two papers are related to the context of sales: one proposed a system on the basis of an artificial neural network (ANN) approach to in store industry to forecast sales [13], one paper proposed a real-time sales management model using agent-based systems (ABSs) [18], and one paper used a genetic algorithm (GA) [2] model to define an online system to benefit sales promotion. One paper used for pricing which utilized AI general approaches for proposing pricing system for different services and products [21]. Two papers focused on simulation

Table 15.2 AI approach categorization on the basis of fields with their frequencies

Field	AI approach
Marketing	1. ANN (artificial neural networks) (2) 2. GA (Genetic Algorithms) (1) 3. Feature Learning (2) 4. Agent-based systems (1) 5. Particle Swarm Optimization (1) 6. Simulation Annealing (2) 7. Association Rules (1) 8. SVM (Support vector machine) (1)
Supply Chain	1. ANN (5) 2. Swarm Intelligence (1) 3. Agent Based Systems (ABSs) (1) 4. Bayesian Networks (1) 5. Data mining (1) 6. SVM (1) 7. Feature Learning (FL) (4) 8. Particle Swarm Optimization (1) 9. Simulation Annealing (1) 10. Association Rules (1)
Production	1. ANN (5) 2. FL Modeling (3) 3. GA (2) 4. Data Mining (1) 5. Agent Based Systems (3)
Logistics	1. ANN (3) 2. Data Mining (1) 3. Simulated Annealing (1) 4. AI General forms (1) 5. Feature Learning (3) 6. Agent based System (2)

annealing; one focused on tree models which combined fuzzy models [29] and cluster models to propose a technique for market segmentation [10]. Various researchers focused primarily on consumer behaviour by the use of an association rule mining model [12] for the suggestion of a consumer-based prediction model; whereas fuzzy system methodology is used to define knowledge discovery in databases in order to support consumer decision behaviours [5, 11]. Researchers [48] explored the SVM approach of AI to contact selection in marketing by carrying the literature review.

As per Table 15.2, a total of 11 articles are assigned to the logistics field of AI contribution in the SCM field. Three papers are based on AI techniques, namely as ANN utilized for management and operations by proposing an intelligent robotics system in this field [18, 41, 42]; researchers propose a multi-dimensional framework based on data mining to categorize between human and artificial collaboration systems in logistics. Researchers identified inter-organizational lot-sizing limitations by the creation of a set of self-interested and autonomous agents by following a feature learning AI approach [9, 39, 45]. Finally, two papers utilize the agent-based systems to evaluate how AI approaches and radio-frequency identification (RFID) can improve the logistics workflow [7, 14].

According to Table 15.2, a total of 14 papers are related to the production field. Two papers worked on the GA methodology for proposing a GA-based model for line-balancing problems in parallel two-side assembly, whereas Sanders worked on AI tools for effective assembly automation [1, 6]. Five papers focused deeply on production forecasting (i.e., petroleum, electricity, machine) by implementing an ANN approach to reduce maintenance costs. Three FL modeling was used in three papers for production planning and scheduling to limit the problems of production management in an integrative manner [39, 45, 49]. One paper focused on identification of knowledge-based system exemplary application by using a data mining AI technique [35]. Three papers focused on approaching efficient manufacturing decision support using case and agent-based systems [8, 15, 51].

17 papers are pertaining to the supply chain field with the majority of papers being related to forecasting. A total of five papers are related to demand forecasting using ANNs and fuzzy techniques in order to manage partial demand with reduced information [30–32, 34, 40]. One paper developed a retailer forecasting model using an AI approach, i.e., Support Vector Machines (SVM), on the basis of customer segmentation to enhance the inventory performance by proposing an intelligent system [43]. One paper focused on supplier selection to build an efficient supplier selection model. By means of one paper on ABSs models in the supply chain, the researchers defined a simulation framework for supply chain planning, defining a negotiation model [25]. A systematic literature review always helps to explore various subfields in the domain of AI for solving SCM problems of practical nature. The researcher in [38] consider the inter-organizational prediction system adoption by using RosettaNet as a tool for case study. Younis et al.

[40] provided an intelligent integrated system for supply chain integration to provide benefits to the AI contribution in the SCM field, context-aware systems, and for reasoning systems.

15.3.3 AI approaches

By "AI Approaches", we mean architectures, algorithms, knowledge or data formalisms, and methodological approaches which can be defined in a clean and precise manner. For the consideration of analysis, we firstly reasoned the scientific sources for reporting the identified list of AI approaches in scientific literature and scientific practice by introducing an association of AI approaches in four subfields of SCM as mentioned in Table 15.2. Table 15.2 defines the AI approaches used in four fields of the literature with their given frequencies during the time span of 2014 to 2022. It is evident from Table 15.2 that most categories in the context of AI approaches lie in the supply chain field with 10 fields due to their practical nature, which mainly considers case studies, experimental research, and real-time problem-solving studies. FL, ANNs, and ABSs are the most widely used approaches in the supply chain literature. With five approaches used, production is the second most frequent field, with ANNs and FL considered the most frequently used approach here. Logistics is the third most diverse field, with 11 articles and 6 AI approaches. ANNs, AI general fields and FL are the most frequent approaches in this field. Finally, marketing is a less varied field, with 8 approaches from 10 research papers. Table 15.2 also represents the final frequency of 52 research papers of AI approaches in literature. It is evident that various studies used double or numerous AI approaches in two ways: by collecting those approaches to constitute a hybrid approach and by implementing in a sequential way. It was also evidently observed from the literature that the most widely used AI approach was ANN (used 15 times) across all the four fields. FL/modeling was the second most used AI approach (9 times) across all the four fields. ABSs was the joint third most used AI approach (used 7 times each). The simulated annealing approach was the next used AI approach (used 4 times), followed by GA, data mining, and SA (used 3 times each), followed by SVM, particle swarm optimization, and association rules (used two times) AI approach. The remaining techniques i.e. Bayesian networks, and swarm intelligence AI approaches, were both used on just one occasion.

15.4 CONCLUSION

The major objective of the present research is to define how AI can potentially contribute to the SCM field on the basis of SLR. We examined 52 papers published which were comprehensible through five phases, as mentioned in the methodology section. The findings of the present study reveal that among the various distinct AI approaches available, some approaches have been implemented in a higher range as compared to other approaches. The most prominent AI approach is ANNs, which aims to determine complex hidden patterns which cannot be found efficiently by human capabilities but ANNs possess solutions in various domains of problems, such as pattern classification, clustering, optimization, process control, prediction, and approximation. The second most used approach is FL (i.e., multiple-valued logic for handling partial truth problems) for building intelligent systems. This is followed by MAS/ABS, which acts as a broader contribution in SCM and works by receiving data from the surrounding atmosphere, and which is followed by acting proactively and autonomously to build a solution for a crucial problem. Other important approaches that can be effectively considered are GAs (which mimics natural selection for tackling several divisions of combinatorial decision problems), data mining (which provides meaningful decisions from huge large data sets by the discovery of data knowledge), swarm intelligence (which mimics social insects' behaviour to provide solutions to complicated problems); and SVMs (which use a linear classifier to classify data). Other, less prominent AI approaches used in SCM studies include simulated annealing, association rule mining, hill climbing, tree-based models, expert systems, k-means clustering, robot programming, heuristics, Bayesian networks, stochastic simulation, Gaussian models and decision tree models. In addition, AI approaches which require further research in the context of industrial adoption in the SCM domain are NLP (machine–human-based interactions), technical specialist (programming, robot dynamics and the optimization techniques for building intelligent robots) etc. Sequentially, in the context of logistics this study reveal that network-based nature can provide a better natural framework to implement with the effective use of AI. In the context of suppliers, the use of effective AI tools which can analyze big data is strongly recommended. Therefore, further research into interactive decision-making systems is required to improve the accuracy of such problems to transform operations of industries from reactive to proactive, manual to autonomous, and the planning of production forecasting to production prediction. In the context of marketing, customer interactions requiring automation is a still potentially promising area. Chatbot or voice defines a new era of customer service, with the capability of increased productivity levels as compared to the human capability for effective customer service enquiries.

15.5 LIMITATIONS AND IMPLICATIONS

As we primarily aimed to achieve a wider analysis of AI in SCM with various subfields, we were not completely able to consider each and every detail of every study. Thus, a single-technique and focused assessment is strongly advised. Since this work has both theoretical and practical implications, it can be efficiently applied to practice and research in the domain of SCM. Firstly, in this study, the most prominent AI approaches applicable in SCM studies are considered to achieve a better insight on the topic. In the context of answering the second SRQ, the prominent AI approaches which can be implemented in SCM research are analyzed. The third SRQ utilized the tasks and subfields of AI in SCM. Finally, in order to answer the final SRQ, the tasks and subfields of AI in SCM which are to be improved by the applications of AI are analyzed for further research studies in the domain. The results of our research study also provides some potential implications for practitioners and managers. For instance, managers can effectively implement GA-based solutions on the arm of bespoke software to consider the existing gaps in the supply chain. In addition to the above, well-planned and oriented AI approaches, namely, as hierarchical planning and distributed problem-solving, are of practical use to SCM managers.

15.6 FUTURE SCOPE

Since some studies have provided potential approaches to the implementation of AI in SC, there are various specific gaps in the current scientific research. For instance, future work could be undertaken in the direction of implementing ABSs with improved complexity management abilities to resolve limitations in supply chain assimilation. It is also observed from the literature that research on real-time pricing (RTP) is moderately country-sensitive, whereas the various studies primarily focused on China. Hence, a future research suggestion topic could be focused on RTP in non-Chinese markets. It is evident from the literature that reverse auctions act as a wider gaps in the context of AI; therefore, further work can be done on reverse auctioning by including supply chain in AI approaches (i.e. heuristic price methodologies). Supply chain and logistics optimization is a very important task. The results of several literature studies focused potentially on developing and creating methodologies, frameworks, models, and solutions. In some cases, some of the studies provide a platform for testing their application, usability, or generalizability in later studies. This gap can be overcome through the use of real-time data for proposed framework testing. In addition, this review work also provides more insights for following future research propositions where every proposition correlates to one or more relevant sub-research questions (SRQs):

1. FL, ANNs, and MAS/ABS are the most predominant AI approaches in SCM. (SRQ 1)
2. FL, ANNs, and MAS/ABS are studied primarily in the SCM systematic literature; therefore, they have had more impact on SCM than the other AI approaches. (SRQ 1,2)

3. To get better insights about the use of AI in SCM, both practitioners and researchers require enthusiastically designed structured problems and software. (SRQ 1,3)
4. It is evident from the literature that empirical studies of AI in SCM have direct and positive impacts on the practical use of AI. (SRQ 1,3,4).

REFERENCES

1. Aleksendrić, D., & Carlone, P. (2015). Soft computing techniques. In D. Aleksendrić, & P. Carlone (Eds.), *Soft Computing in the Design and Manufacturing of Composite Materials*, (vol. 4, pp. 39–60). Oxford: Woodhead Publishing. https://doi.org/10.1533/9781782421801.39
2. Amirkolaii, K. N., Baboli, A., Shahzad, M. K., & Tonadre, R. (2017). Demand forecasting for irregular demands in business aircraft spare parts supply chains by using artificial intelligence (AI). *IFAC-PapersOnLine*, 50, 15221–15226. https://doi.org/10.1016/j.ifacol.2017.08.2371
3. Camarillo, A., Ríos, J., & Althoff, K.-D. (2018). Knowledge-based multiagent system for manufacturing problem solving process in production plants. *Journal of Manufacturing Systems*, 47, 115–127. https://doi.org/10.1016/j. jmsy.2018.04.002
4. Canhoto, A. I., & Clear, F. (2020). Artificial intelligence and machine learning as business tools: A framework for diagnosing value destruction potential. *Business Horizons*, 63, 183–193. https://doi.org/10.1016/j. bushor.2019. 11.003
5. Casabayó, M., Agell, N., & Sanchez-Hernández, G. (2015). Improved market segmentation by fuzzifying crisp clusters: A case study of the energy market in Spain. *Expert Systems with Applications*, 42, 1637–1643. https://doi. org/10.1016/j. eswa.2014.09.044
6. Chong, A. Y.-L., & Bai, R. (2014). Predicting open IOS adoption in SMEs: An integrated SEM-neural network approach. *Expert Systems with Applications*, 41, 221–229. https://doi.org/10.1016/j.eswa.2013.07.023
7. Dirican, C. (2015). The impacts of robotics, artificial intelligence on business and economics. *Procedia Social and Behavioral Sciences*, 195, 564–573. https://doi.org/10.1016/j.sbspro.2015.06.134
8. Dubey, R., Gunasekaran, A., Childe, S. J., Bryde, D. J., Giannakis, M., Foropon, C., ... Hazen, B. T. (2020). Big data analytics and artificial intelligence pathway to operational performance under the effects of entrepreneurial orientation and environmental dynamism: A study of manufacturing organisations. *International Journal of Production Economics*, 226, Article 107599. https:// doi.org/10.1016/j. ijpe.2019.107599
9. Ellram, L. M., & Ueltschy Murfield, M. L. (2019). Supply chain management in industrial marketing–Relationships matter. *Industrial Marketing Management*, 79, 36–45. https://doi.org/10.1016/j.indmarman.2019.03.007
10. Ennen, P., Reuter, S., Vossen, R., & Jeschke, S. (2016). Automated production ramp-up through self-learning systems. *Procedia CIRP*, 51, 57–62. https://doi. org/10.1016/j.procir.2016.05.094

11. Eslikizi, S., Ziebuhr, M., Kopfer, H., & Buer, T. (2015). Shapley-based side payments and simulated annealing for distributed lot-sizing. *IFAC-PapersOnLine*, 48, 1592–1597. https://doi.org/10.1016/j.ifacol.2015.06.313

12. Heger, J., Branke, J., Hildebrandt, T., & Scholz-Reiter, B. (2016). Dynamic adjustment of dispatching rule parameters in flow shops with sequence-dependent set-up times. *International Journal of Production Research*, 54, 6812–6824. https://doi.org/10.1080/00207543.2016.1178406

13. Hossein Javaheri, S., Mehdi Sepehri, M., & Teimourpour, B. (2014). Response modeling in direct marketing: A data mining based approach for target selection. *Data Mining Applications with R*, 153–178. https://doi.org/10.1016/B978-0-12-411511-8.00006-2

14. Jarrahi, M. H. (2018). Artificial intelligence and the future of work: Human-AI symbiosis in organizational decision making. *Business Horizons*, 61, 577–586. https://doi.org/10.1016/j.bushor.2018.03.007

15. Kaplan, A., & Haenlein, M. (2020). Rulers of the world, unite! The challenges and opportunities of artificial intelligence. *Business Horizons*, 63, 37–50. https://doi.org/10.1016/j.bushor.2019.09.003

16. Kasie, F. M., Bright, G., & Walker, A. (2017). Decision support systems in manufacturing: A survey and future trends. *Journal of Modelling in Management*. https://doi.org/10.1108/JM2-02-2016-0015

17. Klumpp, M. (2018). Automation and artificial intelligence in business logistics systems: Human reactions and collaboration requirements. *International Journal of Logistics Research and Applications*, 21, 224–242. https://doi.org/10.1080/13675567.2017.1384451

18. Knoll, D., Prüglmeier, M., & Reinhart, G. (2016). Predicting future inbound logistics processes using machine learning. *Procedia CIRP*, 52, 145–150. https://doi.org/10.1016/j.procir.2016.07.078

19. Kohtamäki, M., Parida, V., Oghazi, P., Gebauer, H., & Baines, T. (2019). Digital servitization business models in ecosystems: A theory of the firm. *Journal of Business Research*, 104, 380–392. https://doi.org/10.1016/j.jbusres.2019.06.027

20. Kotler, P., Manrai, L. A., Lascu, D.-N., & Manrai, A. K. (2019). Influence of country and company characteristics on international business decisions: A review, conceptual model, and propositions. *International Business Review*, 28, 482–498. https://doi.org/10.1016/j.ibusrev.2018.11.006

21. Kumar, V., Ramachandran, D., & Kumar, B. (2020). Influence of new-age technologies on marketing: A research agenda. *Journal of Business Research*. https://doi.org/10.1016/j.jbusres.2020.01.007

22. Kwong, C. K., Jiang, H., & Luo, X. G. (2016). AI-based methodology of integrating affective design, engineering, and marketing for defining design specifications of new products. *Engineering Applications of Artificial Intelligence*, 47, 49–60. https://doi.org/10.1016/j.engappai.2015.04.001

23. Ładyżyński, P., Żbikowski, K., & Gawrysiak, P. (2019). Direct marketing campaigns in retail banking with the use of deep learning and random forests. *Expert Systems with Applications*, 134, 28–35. https://doi.org/10.1016/j.eswa.2019.05.020

24. Mayr, A., Weigelt, M., Masuch, M., Meiners, M., Hüttel, F., & Franke, J. (2018). Application scenarios of artificial intelligence in electric drives production. *Procedia Manufacturing*, 24, 40–47. https://doi.org/10.1016/j.promfg.2018.06.006

25. Merlino, M., & Sproģe, I. (2017). The augmented supply chain. *Procedia Engineering*, 178, 308–318. https://doi.org/10.1016/j.proeng.2017.01.053

26. Min, H. (2015). Genetic algorithm for supply chain modelling: Basic concepts and applications. *International Journal of Services and Operations Management*, 22, 143–163. https://doi.org/10.1504/IJSOM.2015.071527

27. Mobarakeh, N. A., Shahzad, M. K., Baboli, A., & Tonadre, R. (2017). Improved forecasts for uncertain and unpredictable spare parts demand in business aircraft's with bootstrap method. *IFAC-PapersOnLine*, 50, 15241–15246. https://doi.org/10.1016/j. ifacol.2017.08.2379

28. Regal, T., & Pereira, C. E. (2018). Ontology for conceptual modelling of intelligent maintenance systems and spare parts supply chain integration. *IFAC-PapersOnLine*, 51, 1511–1516. https://doi.org/10.1016/j.ifacol.2018.08. 285

29. Rekha, A. G., Abdulla, M. S., & Asharaf, S. (2016). Artificial intelligence marketing: An application of a novel lightly trained support vector data description. *Journal of Information and Optimization Sciences*, 37, 681–691. https://doi.org/10.1080/02522667.2016.1191186

30. Toorajipour, R., Sohrabpour, V., Nazarpour, A., Oghazi, P., & Fischl, M. (2021). Artificial intelligence in supply chain management: A systematic literature review. *Journal of Business Research*, 122, 502–517. https://doi.org/10.1016/j. jbusres.2020.09.009

31. Soni, N., Sharma, E. K., Singh, N., & Kapoor, A. (2020). Artificial intelligence in business: From research and innovation to market deployment. *Procedia Computer Science. International Conference on Computational Intelligence and Data Science*, 167, 2200–2210. https://doi.org/10.1016/j. procs.2020.03.272

32. Stalidis, G., Karapistolis, D., & Vafeiadis, A. (2015). Marketing decision support using artificial intelligence and knowledge modeling: Application to tourist destination management. *Procedia – Social and Behavioral Sciences*, 175, 106–113. https://doi.org/10.1016/j.sbspro.2015.01.1180

33. Ting, S. L., Tse, Y. K., Ho, G. T. S., Chung, S. H., & Pang, G. (2014). Mining logistics data to assure the quality in a sustainable food supply chain: A case in the red wine industry. *International Journal of Production Economics*, 152, 200–209.

34. Townsend, D. M., & Hunt, R. A. (2019). Entrepreneurial action, creativity, & judgment in the age of artificial intelligence. *Journal of Business Venturing Insights*, 11, Article e00126. https://doi.org/10.1016/j.jbvi.2019.e00126

35. Wang, J., & Yue, H. (2017). Food safety pre-warning system based on data mining for a sustainable food supply chain. *Food Control*, 73, 223–229. https://doi.org/10.1016/j.foodcont.2016.09.048

36. Zhang, X., Chan, F. T. S., Adamatzky, A., Mahadevan, S., Yang, H., Zhang, Z., & Deng, Y. (2017). An intelligent physarum solver for supply chain network design under profit maximization and oligopolistic competition. *International Journal of Production Research*, 55, 244–263. https://doi.org/10.1080/002075 43.2016.1203075

37. Dhar, V., Geva, T., Oestreicher-Singer, G., & Sundararajan, A. (2014). Prediction in economic networks. *Information Systems Research*, 25(2), 264–284. http://www.jstor.org/stable/24700173

38. Sanders, R. N., Boone, T., Ganeshan, R., & Wood, D. J. (2019). Sustainable supply chain in the age of AI and digitization: Research challenges and opportunities. *Journal of Business Logistics*, 40(3), 229–240.

39. Dhamija, P., & Bag, S. (2020). Role of artificial intelligence in operations environment: A review and bibliometric analysis. *The TQM Journal*, 32(4), 869–896. https://doi.org/10.1108/TQM-10-2019-0243

40. Younis, H., Sundarakani, B., & Alsharairi, M. (2022). Applications of artificial intelligence and machine learning within supply chains: Systematic review and future research directions. *Journal of Modelling in Management*, 17(3), 916–940. https://doi.org/10.1108/JM2-12-2020-0322

41. Akbari, M., & Do, T.N.A. (2021). A systematic review of machine learning in logistics and supply chain management: Current trends and future directions. *Benchmarking: An International Journal*, 28(10), 2977–3005. https://doi.org/10.1108/BIJ-10-2020-0514

42. Pålsson, H., & Sandberg, E. (2022). Packaging paradoxes in food supply chains: Exploring characteristics, underlying reasons and management strategies. *International Journal of Physical Distribution & Logistics Management*, 52(11), 25–52. https://doi.org/10.1108/IJPDLM-09-2019-0270

43. Zhang, H., Shi, Y., Yang, X., & Zhou, R. (2021). A firefly algorithm modified support vector machine for the credit risk assessment of supply chain finance. *Research in International Business and Finance*, 58. https://doi.org/10.1016/j.ribaf.2021.101482

44. Lin, H., Lin, J., & Wang, F., (2022). An innovative machine learning model for supply chain management. *Journal of Innovation & Knowledge*, 7(4). https://doi.org/10.1016/j.jik.2022.100276

45. Liu, N., Bouzembrak, Y., van den Bulk, L. M., Gavai, A., van den Heuvel, L. J., & Marvin, H. J. P. (2022). Automated food safety early warning system in the dairy supply chain using machine learning. *Journal of Food Control*, 136. https://doi.org/10.1016/j.foodcont.2022.108872

46. Toorajipour, R., Sohrabpour, V., Nazarpour, A., Oghazi, P., & Fischl, M. (2021). Artificial intelligence in supply chain management: A systematic literature review. *Journal of Business Research*, 122, 502–517. https://doi.org/10.1016/j.jbusres.2020.09.009

47. Kosasih, E. E., & Brintrup, A. (2022). A machine learning approach for predicting hidden links in supply chain with graph neural networks. *International Journal of Production Research*, 60(17), 5380–5393. https://doi.org/10.1080/00207543.2021.1956697

48. Chhajer, P., Shah, M., & Kshirsagar, A. (2022). The applications of artificial neural networks, support vector machines, and long–short term memory for stock market prediction. *Decision Analytics Journal*, 2. https://doi.org/10.1016/j.dajour.2021.100015

49. Safaei, S., Ghasemi, P., Goodarzian, F., & Momenitabar, M. (2022). Designing a new multi-echelon multi-period closed-loop supply chain network by forecasting demand using time series model: A genetic algorithm. *Journal of Environmental science and Pollution Research*, 29, 79754–79768.

50. Mastos, T. D., Nizamis, A., Terzi, S., Gkortzis, D., Papadopoulos, A., Tsagkalidis, N., Ioannidis, D., Votis, K., Tzovaras, D., (2021). Introducing an application of an industry 4.0 solution for circular supply chain management. *Journal of Cleaner Production*, 300. https://doi.org/10.1016/j.jclepro.2021.126886

51. Aslam, J., & Saleem, A, Khan, N. T., Kim, Y. B. (2021). Factors influencing blockchain adoption in supply chain management practices: A study based on the oil industry. *Journal of Innovation & Knowledge*, 6(2), 124–134. https://doi.org/10.1016/j.jik.2021.01.002

52. Birkel, H., & Müller, J. M. (2021). Potentials of industry 4.0 for supply chain management within the triple bottom line of sustainability – A systematic literature review. *Journal of Cleaner Production*, 289. https://doi.org/10.1016/j.jclepro.2020.125612

Index

Pages in *italics* refer to figures and pages in **bold** refer to tables.

Milton Keynes UK
Ingram Content Group UK Ltd.
UKHW031129141024
449569UK00006B/331

9 781032 426730